Ratner's Theorems on Unipotent Flows

CHICAGO LECTURES IN MATHEMATICS SERIES
Editors: Spencer J. Bloch, Peter Constantin, Benson Farb,
Norman R. Lebovitz, Carlos Kenig, and J. P. May

Other *Chicago Lectures in Mathematics* titles available from
the University of Chicago Press

RATNER'S THEOREMS ON UNIPOTENT FLOWS

Dave Witte Morris

THE UNIVERSITY OF CHICAGO PRESS • CHICAGO AND LONDON

Dave Witte Morris is professor of mathematics at the University of Lethbridge.

The University of Chicago Press, Chicago 60637
The University of Chicago Press, Ltd., London
© 2005 by Dave Witte Morris
All rights reserved. Published 2005
Printed in the United States of America

14 13 12 11 10 09 08 07 06 05   1 2 3 4 5

ISBN: 0-226-53983-0   (cloth)
ISBN: 0-226-53984-9   (paper)

Library of Congress Cataloging-in-Publication Data
Morris, Dave Witte.
  Ratner's theorems on unipotent flows / Dave Witte Morris.
      p. cm. — (Chicago lectures in mathematics series)
  Includes bibliographical references and index.
  ISBN 0-226-53983-0 (alk. paper) — ISBN 0-226-53984-9 (pbk. : alk. paper)
  1. Ergodic theory. 2. Measure theory. 3. Number theory. 4. Lie groups. 5. Diophantine
equations. I. Title. II. Chicago lectures in mathematics.
  QA611.5.M67 2005
  515'.48—dc22

                                                              2005005826

⊚ The paper used in this publication meets the minimum requirements of the American
National Standard for Information Sciences—Permanence of Paper for Printed Library
Materials, ANSI Z39.48-1992.

To Joy,
my wife and friend

Contents

Abstract

Unipotent flows are well-behaved dynamical systems. In particular, Marina Ratner has shown that the closure of every orbit for such a flow is of a nice algebraic (or geometric) form. This is known as the Ratner Orbit Closure Theorem; the Ratner Measure-Classification Theorem and the Ratner Equidistribution Theorem are closely related results. After presenting these important theorems and some of their consequences, the lectures explain the main ideas of the proof. Some algebraic technicalities will be pushed to the background.

Chapter 1 is the main part of the book. It is intended for a fairly general audience, and provides an elementary introduction to the subject, by presenting examples that illustrate the theorems, some of their applications, and the main ideas involved in the proof.

Chapter 2 gives an elementary introduction to the theory of entropy, and proves an estimate used in the proof of Ratner's Theorems. It is of independent interest.

Chapters 3 and 4 are utilitarian. They present some basic facts of ergodic theory and the theory of algebraic groups that are needed in the proof. The reader (or lecturer) may wish to skip over them, and refer back as necessary.

Chapter 5 presents a fairly complete (but not entirely rigorous) proof of Ratner's Measure-Classification Theorem. Unlike the other chapters, it is rather technical. The entropy argument that finishes our presentation of the proof is due to G. A. Margulis and G. Tomanov. Earlier parts of our argument combine ideas from Ratner's original proof with the approach of G. A. Margulis and G. Tomanov.

The first four chapters can be read independently, and are intended to be largely accessible to second-year graduate students. All four are needed for Chapter 5. A reader who is familiar with ergodic theory and algebraic groups, but not unipotent flows, may skip Chaps. 2, 3, and 4 entirely, and read only §1.5–§1.8 of Chap. 1 before beginning Chap. 5.

Possible lecture schedules

It is quite reasonable to stop anywhere after §1.5. In particular, a single lecture (1–2 hours) can cover the main points of §1.1–§1.5.

A good selection for a moderate series of lectures would be §1.1–§1.8 and §5.1, adding §2.1–§2.5 if the audience is not familiar with entropy. For a more logical presentation, one should briefly discuss §3.1 (the Pointwise Ergodic Theorem) before starting §1.5–§1.8.

Here are suggested guidelines for a longer course:

§1.1–§1.3: Introduction to Ratner's Theorems (0.5–1.5 hours)
§1.4: Applications of Ratner's Theorems (optional, 0–1 hour)
§1.5–§1.6: Shearing and polynomial divergence (1–2 hours)
§1.7–§1.8: Other basic ingredients of the proof (1–2 hours)
§1.9: From measures to orbit closures (optional, 0–1 hour)

§2.1–§2.3: What is entropy? (1–1.5 hours)
§2.4–§2.5: How to calculate entropy (1–2 hours)
§2.6: Proof of the entropy estimate (optional, 1–2 hours)

§3.1: Pointwise Ergodic Theorem (0.5–1.5 hours)
§3.2: Mautner Phenomenon (optional, 0.5–1.5 hours)
§3.3: Ergodic decomposition (optional, 0.5–1.5 hours)
§3.4: Averaging sets (0.5–1.5 hours)

§4.1–§4.9: Algebraic groups (optional, 0.5–3 hours)

§5.1: Outline of the proof (0.5–1.5 hours)
§5.2–§5.7: A fairly complete proof (3–5 hours)
§5.8–§5.9: Making the proof more rigorous (optional, 1–3 hours)

Acknowledgments

I owe a great debt to many people, including the audiences of my lectures, for their many comments and helpful discussions that added to my understanding of this material and improved its presentation. A few of the main contributors are S. G. Dani, Alex Eskin, Bassam Fayad, David Fisher, Alex Furman, Elon Lindenstrauss, G. A. Margulis, Howard Masur, Marina Ratner, Nimish Shah, Robert J. Zimmer, and three anonymous referees. I am also grateful to Michael Koplow, for pointing out many, many minor errors in the manuscript.

Major parts of the book were written while I was visiting the Tata Institute of Fundamental Research (Mumbai, India), the Federal Technical Institute (ETH) of Zurich, and the University of Chicago. I am grateful to my colleagues at all three of these institutions for their hospitality and for the aid they gave me in my work on this project. Some financial support was provided by a research grant from the National Science Foundation (DMS-0100438).

I gave a series of lectures on this material at the ETH of Zurich and at the University of Chicago. Chapter 1 is an expanded version of a lecture that was first given at Williams College in 1990, and has been repeated at several other universities. Chapter 2 is based on talks for the University of Chicago's Analysis Proseminar in 1984 and Oklahoma State University's Lie Groups Seminar in 2002.

I thank my wife, Joy Morris, for her emotional support and unfailing patience during the writing of this book.

All author royalties from sales of this book will go to charity.

CHAPTER 1

Introduction to Ratner's Theorems

1.1. What is Ratner's Orbit Closure Theorem?

We begin by looking at an elementary example.

(1.1.1) Example. For convenience, let us use $[x]$ to denote the image of a point $x \in \mathbb{R}^n$ in the n-torus $\mathbb{T}^n = \mathbb{R}^n/\mathbb{Z}^n$; that is,

$$[x] = x + \mathbb{Z}^n.$$

Any vector $v \in \mathbb{R}^n$ determines a C^∞ flow φ_t on $\mathbb{T}^n$, by

$$\varphi_t([x]) = [x + tv] \qquad \text{for } x \in \mathbb{R}^n \text{ and } t \in \mathbb{R} \qquad (1.1.2)$$

(see Exer. 2). It is well known that the closure of the orbit of each point of $\mathbb{T}^n$ is a subtorus of $\mathbb{T}^n$ (see Exer. 5, or see Exers. 3 and 4 for examples). More precisely, for each $x \in \mathbb{R}^n$, there is a vector subspace S of $\mathbb{R}^n$, such that

S1) $v \in S$ (so the entire φ_t-orbit of $[x]$ is contained in $[x+S]$),

S2) the image $[x + S]$ of $x + S$ in $\mathbb{T}^n$ is compact (hence, the image is diffeomorphic to $\mathbb{T}^k$, for some $k \in \{0, 1, 2, \ldots, n\}$), and

S3) the φ_t-orbit of $[x]$ is dense in $[x + S]$ (so $[x + S]$ is the closure of the orbit of $[x]$).

In short, the closure of every orbit is a nice, geometric subset of $\mathbb{T}^n$.

Ratner's Orbit Closure Theorem is a far-reaching generalization of Eg. 1.1.1. Let us examine the building blocks of that example.

- Note that $\mathbb{R}^n$ is a *Lie group*. That is, it is a group (under vector addition) and a manifold, and the group operations are C^∞ functions.

- The subgroup $\mathbb{Z}^n$ is *discrete*. (That is, it has no accumulation points.) Therefore, the quotient space $\mathbb{R}^n/\mathbb{Z}^n = \mathbb{T}^n$ is a manifold.

- The quotient space $\mathbb{R}^n/\mathbb{Z}^n$ is compact.

1

- The map $t \mapsto tv$ (which appears in the formula (1.1.2)) is a **one-parameter subgroup** of $\mathbb{R}^n$; that is, it is a C^∞ group homomorphism from $\mathbb{R}$ to $\mathbb{R}^n$.

Ratner's Theorem allows:

- the Euclidean space $\mathbb{R}^n$ to be replaced by any **Lie group** G;
- the subgroup $\mathbb{Z}^n$ to be replaced by any **discrete subgroup** Γ of G, such that the quotient space $\Gamma \backslash G$ is compact; and
- the map $t \mapsto tv$ to be replaced by any **unipotent** one-parameter subgroup u^t of G. (The definition of "unipotent" will be explained later.)

Given G, Γ, and u^t, we may define a C^∞ flow φ_t on $\Gamma \backslash G$ by

$$\varphi_t(\Gamma x) = \Gamma x u^t \qquad \text{for } x \in G \text{ and } t \in \mathbb{R} \qquad (1.1.3)$$

(cf. 1.1.2 and see Exer. 7). We may also refer to φ_t as the u^t-**flow** on $\Gamma \backslash G$. Ratner proved that the closure of every φ_t-orbit is a nice, geometric subset of $\Gamma \backslash G$. More precisely (note the direct analogy with the conclusions of Eg. 1.1.1), if we write $[x]$ for the image of x in $\Gamma \backslash G$, then, for each $x \in G$, there is a closed, connected subgroup S of G, such that

S1') $\{u^t\}_{t \in \mathbb{R}} \subset S$ (so the entire φ_t-orbit of $[x]$ is contained in $[xS]$),

S2') the image $[xS]$ of xS in $\Gamma \backslash G$ is compact (hence, diffeomorphic to the homogeneous space $\Lambda \backslash S$, for some discrete subgroup Λ of S), and

S3') the φ_t-orbit of $[x]$ is dense in $[xS]$ (so $[xS]$ is the closure of the orbit).

(1.1.4) Remark.

1) Recall that $\Gamma \backslash G = \{ \Gamma x \mid x \in G \}$ is the set of **right** cosets of Γ in G. We will consistently use right cosets Γx, but all of the results can easily be translated into the language of **left** cosets $x\Gamma$. For example, a C^∞ flow φ'_t can be defined on G/Γ by $\varphi'_t(x\Gamma) = u^t x \Gamma$.

2) It makes no difference whether we write $\mathbb{R}^n / \mathbb{Z}^n$ or $\mathbb{Z}^n \backslash \mathbb{R}^n$ for $\mathbb{T}^n$, because right cosets and left cosets are the same in an abelian group.

(1.1.5) Notation. For a very interesting special case, which will be the main topic of most of this chapter,

- let

$$G = \mathrm{SL}(2, \mathbb{R})$$

be the group of 2×2 real matrices of determinant one; that is

$$SL(2, \mathbb{R}) = \left\{ \begin{bmatrix} a & b \\ c & d \end{bmatrix} \,\middle|\, \begin{array}{l} a, b, c, d \in \mathbb{R}, \\ ad - bc = 1 \end{array} \right\},$$

and

- define $u, a \colon \mathbb{R} \to SL(2, \mathbb{R})$ by

$$u^t = \begin{bmatrix} 1 & 0 \\ t & 1 \end{bmatrix} \quad \text{and} \quad a^t = \begin{bmatrix} e^t & 0 \\ 0 & e^{-t} \end{bmatrix}.$$

Easy calculations show that

$$u^{s+t} = u^s u^t \quad \text{and} \quad a^{s+t} = a^s a^t$$

(see Exer. 8), so u^t and a^t are one-parameter subgroups of G. For any subgroup Γ of G, define flows η_t and γ_t on $\Gamma \backslash G$, by

$$\eta_t(\Gamma x) = \Gamma x u^t \quad \text{and} \quad \gamma_t(\Gamma x) = \Gamma x a^t.$$

(1.1.6) Remark. Assume (as usual) that Γ is **discrete** and that $\Gamma \backslash G$ is compact. If $G = SL(2, \mathbb{R})$, then, in geometric terms,

1) $\Gamma \backslash G$ is (essentially) the unit tangent bundle of a compact surface of constant negative curvature (see Exer. 10),

2) γ_t is called the **geodesic flow** on $\Gamma \backslash G$ (see Exer. 11), and

3) η_t is called the **horocycle flow** on $\Gamma \backslash G$ (see Exer. 11).

(1.1.7) Definition. A square matrix T is **unipotent** if 1 is the only (complex) eigenvalue of T; in other words, $(T - 1)^n = 0$, where n is the number of rows (or columns) of T.

(1.1.8) Example. Because u^t is a unipotent matrix for every t, we say that u^t is a **unipotent one-parameter subgroup** of G. Thus, Ratner's Theorem applies to the horocycle flow η_t: the closure of every η_t-orbit is a nice, geometric subset of $\Gamma \backslash G$.

More precisely, algebraic calculations, using properties (S1′, S2′, S3′) show that $S = G$ (see Exer. 13). Thus, the closure of every orbit is $[G] = \Gamma \backslash G$. In other words, every η_t-orbit is dense in the entire space $\Gamma \backslash G$.

(1.1.9) Counterexample. In contrast, a^t is **not** a unipotent matrix (unless $t = 0$), so $\{a^t\}$ is **not** a unipotent one-parameter subgroup. Therefore, Ratner's Theorem does **not** apply to the geodesic flow γ_t.

Indeed, although we omit the proof, it can be shown that the closures of some orbits of γ_t are very far from being nice, geometric subsets of $\Gamma \backslash G$. For example, the closures of some orbits are fractals (nowhere close to being a submanifold of $\Gamma \backslash G$).

Specifically, for some orbits, if C is the closure of the orbit, then some neighborhood (in C) of a point in C is homeomorphic to $C' \times \mathbb{R}$, where C' is a Cantor set.

When we discuss some ideas of Ratner's proof (in §1.5), we will see, more clearly, why the flow generated by this diagonal one-parameter subgroup behaves so differently from a unipotent flow.

(1.1.10) **Remark.** It can be shown fairly easily that *almost* every orbit of the horocycle flow η_t is dense in $[G]$, and the same is true for the geodesic flow γ_t (cf. 3.2.7 and 3.2.4). Thus, for both of these flows, it is easy to see that the closure of *almost every* orbit is $[G]$, which is certainly a nice manifold. (This means that the fractal orbits of (1.1.9) are exceptional; they form a set of measure zero.) The point of Ratner's Theorem is that it replaces "almost every" by "every."

Our assumption that $\Gamma \backslash G$ is compact can be relaxed.

(1.1.11) **Definition.** Let Γ be a subgroup of a Lie group G.

- A measure μ on G is **left invariant** if $\mu(gA) = \mu(A)$ for all $g \in G$ and all measurable $A \subset G$. Similarly, μ is **right invariant** if $\mu(Ag) = \mu(A)$ for all g and A.
- Recall that any Lie group G has a (left) **Haar measure**; that is, there exists a left-invariant (regular) Borel measure μ on G. Furthermore, μ is unique up to a scalar multiple. (There is also a measure that is right invariant, but the right-invariant measure may not be the same as the left-invariant measure.)
- A **fundamental domain** for a subgroup Γ of a group G is a measurable subset $\mathcal{F}$ of G, such that
  ○ $\Gamma\mathcal{F} = G$, and
  ○ $\gamma\mathcal{F} \cap \mathcal{F}$ has measure 0, for all $\gamma \in \Gamma \smallsetminus \{e\}$.
- A subgroup Γ of a Lie group G is a **lattice** if
  ○ Γ is discrete, and
  ○ some (hence, every) fundamental domain for Γ has finite measure (see Exer. 14).

(1.1.12) **Definition.** If Γ is a lattice in G, then there is a unique G-invariant probability measure μ_G on $\Gamma \backslash G$ (see Exers. 15, 16, and 17). It turns out that μ_G can be represented by a smooth volume form on the manifold $\Gamma \backslash G$. Thus, we may say that $\Gamma \backslash G$ has *finite volume*. We often refer to μ_G as the **Haar measure** on $\Gamma \backslash G$.

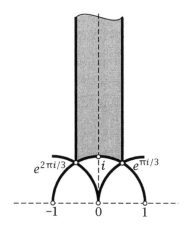

FIGURE 1.1A. When $SL(2, \mathbb{R})$ is identified with (a double cover of the unit tangent bundle of) the upper half plane $\mathfrak{H}$, the shaded region is a fundamental domain for $SL(2, \mathbb{Z})$.

(1.1.13) **Example.** Let
- $G = SL(2, \mathbb{R})$ and
- $\Gamma = SL(2, \mathbb{Z})$.

It is well known that Γ is a lattice in G. For example, a fundamental domain $\mathcal{F}$ is illustrated in Fig. 1.1A (see Exer. 18), and an easy calculation shows that the (hyperbolic) measure of this set is finite (see Exer. 19).

Because compact sets have finite measure, one sees that if $\Gamma \backslash G$ is compact (and Γ is **discrete**!), then Γ is a lattice in G (see Exer. 21). Thus, the following result generalizes our earlier description of Ratner's Theorem. Note, however, that the subspace $[xS]$ may no longer be compact; it, too, may be a noncompact space of finite volume.

(1.1.14) **Theorem** (Ratner Orbit Closure Theorem). *If*
- *G is any Lie group,*
- *Γ is any lattice in G, and*
- *φ_t is any unipotent flow on $\Gamma \backslash G$,*

then the closure of every φ_t-orbit is homogeneous.

(1.1.15) **Remark.** Here is a more precise statement of the conclusion of Ratner's Theorem (1.1.14).
- Use $[x]$ to denote the image in $\Gamma \backslash G$ of an element x of G.

- Let u^t be the unipotent one-parameter subgroup corresponding to φ_t, so $\varphi_t([x]) = [\Gamma x u^t]$.

Then, for each $x \in G$, there is a connected, closed subgroup S of G, such that

1) $\{u^t\}_{t \in \mathbb{R}} \subset S$,

2) the image $[xS]$ of xS in $\Gamma \backslash G$ is closed, and has finite S-invariant volume (in other words, $(x^{-1}\Gamma x) \cap S$ is a lattice in S (see Exer. 22)), and

3) the φ_t-orbit of $[x]$ is dense in $[xS]$.

(1.1.16) Example.

- Let $G = \mathrm{SL}(2, \mathbb{R})$ and $\Gamma = \mathrm{SL}(2, \mathbb{Z})$ as in Eg. 1.1.13.
- Let u^t be the usual unipotent one-parameter subgroup of G (as in Notn. 1.1.5).

Algebraists have classified all of the connected subgroups of G that contain u^t. They are:

1) $\{u^t\}$,

2) the lower-triangular group $\left\{ \begin{bmatrix} * & 0 \\ * & * \end{bmatrix} \right\}$, and

3) G.

It turns out that the lower-triangular group does not have a lattice (cf. Exer. 13), so we conclude that the subgroup S must be either $\{u^t\}$ or G.

In other words, we have the following dichotomy:

each orbit of the u^t-flow on $\mathrm{SL}(2, \mathbb{Z}) \backslash \mathrm{SL}(2, \mathbb{R})$
is either closed or dense.

(1.1.17) Example. Let

- $G = \mathrm{SL}(3, \mathbb{R})$,
- $\Gamma = \mathrm{SL}(3, \mathbb{Z})$, and
- $u^t = \begin{bmatrix} 1 & 0 & 0 \\ t & 1 & 0 \\ 0 & 0 & 1 \end{bmatrix}$.

Some orbits of the u^t-flow are closed, and some are dense, but there are also intermediate possibilities. For example, $\mathrm{SL}(2, \mathbb{R})$ can be embedded in the top left corner of $\mathrm{SL}(3, \mathbb{R})$:

$$\mathrm{SL}(2, \mathbb{R}) \cong \left\{ \begin{bmatrix} * & * & 0 \\ * & * & 0 \\ 0 & 0 & 1 \end{bmatrix} \right\} \subset \mathrm{SL}(3, \mathbb{R}).$$

This induces an embedding

$$\mathrm{SL}(2, \mathbb{Z}) \backslash \mathrm{SL}(2, \mathbb{R}) \hookrightarrow \mathrm{SL}(3, \mathbb{Z}) \backslash \mathrm{SL}(3, \mathbb{R}). \qquad (1.1.18)$$

The image of this embedding is a submanifold, and it is the closure of certain orbits of the u^t-flow (see Exer. 25).

(1.1.19) **Remark.** Ratner's Theorem (1.1.14) also applies, more generally, to the orbits of any subgroup H that is generated by unipotent elements, not just a one-dimensional subgroup. (However, if the subgroup is disconnected, then the subgroup S of Rem. 1.1.15 may also be disconnected. It is true, though, that every connected component of S contains an element of H.)

Exercises for §1.1.

#1. Show that, in general, the closure of a submanifold may be a bad set, such as a fractal. (Ratner's Theorem shows that this pathology cannot not appear if the submanifold is an orbit of a "unipotent" flow.) More precisely, for any closed subset C of $\mathbb{T}^2$, show there is an injective C^∞ function $f: \mathbb{R} \to \mathbb{T}^3$, such that

$$\overline{f(\mathbb{R})} \cap (\mathbb{T}^2 \times \{0\}) = C \times \{0\},$$

where $\overline{f(\mathbb{R})}$ denotes the closure of the image of f.
[*Hint:* Choose a countable, dense subset $\{c_n\}_{n=-\infty}^{\infty}$ of C, and choose f (carefully!) with $f(n) = c_n$.]

#2. Show that (1.1.2) defines a C^∞ **flow** on $\mathbb{T}^n$; that is,

(a) φ_0 is the identity map,

(b) φ_{s+t} is equal to the composition $\varphi_s \circ \varphi_t$, for all $s, t \in \mathbb{R}$; and

(c) the map $\varphi: \mathbb{T}^n \times \mathbb{R} \to \mathbb{T}^n$, defined by $\varphi(x, t) = \varphi_t(x)$ is C^∞.

#3. Let $v = (\alpha, \beta) \in \mathbb{R}^2$. Show, for each $x \in \mathbb{R}^2$, that the closure of $[x + \mathbb{R}v]$ is

$$\begin{cases} \{[x]\} & \text{if } \alpha = \beta = 0, \\ [x + \mathbb{R}v] & \text{if } \alpha/\beta \in \mathbb{Q} \text{ (or } \beta = 0), \\ \mathbb{T}^2 & \text{if } \alpha/\beta \notin \mathbb{Q} \text{ (and } \beta \neq 0). \end{cases}$$

#4. Let $v = (\alpha, 1, 0) \in \mathbb{R}^3$, with α irrational, and let φ_t be the corresponding flow on $\mathbb{T}^3$ (see 1.1.2). Show that the subtorus $\mathbb{T}^2 \times \{0\}$ of $\mathbb{T}^3$ is the closure of the φ_t-orbit of $(0, 0, 0)$.

#5. Given x and v in $\mathbb{R}^n$, show that there is a vector subspace S of $\mathbb{R}^n$, that satisfies (S1), (S3), and (S3) of Eg. 1.1.1.

#6. Show that the subspace S of Exer. 5 depends only on v, not on x. (This is a special property of abelian groups; the analogous statement is **not** true in the general setting of Ratner's Theorem.)

#7. Given
- a Lie group G,
- a closed subgroup Γ of G, and
- a one-parameter subgroup g^t of G,

show that $\varphi_t(\Gamma x) = \Gamma x g^t$ defines a flow on $\Gamma \backslash G$.

#8. For u^t and a^t as in Notn. 1.1.5, and all $s, t \in \mathbb{R}$, show that
(a) $u^{s+t} = u^s u^t$, and
(b) $a^{s+t} = a^s a^t$.

#9. Show that the subgroup $\{a^s\}$ of $SL(2, \mathbb{R})$ normalizes the subgroup $\{u^t\}$. That is, $a^{-s}\{u^t\}a^s = \{u^t\}$ for all s.

#10. Let $\mathfrak{H} = \{x + iy \in \mathbb{C} \mid y > 0\}$ be the **upper half plane** (or **hyperbolic plane**), with Riemannian metric $\langle \cdot \mid \cdot \rangle$ defined by

$$\langle v \mid w \rangle_{x+iy} = \frac{1}{y^2}(v \cdot w),$$

for tangent vectors $v, w \in T_{x+iy}\mathfrak{H}$, where $v \cdot w$ is the usual Euclidean inner product on $\mathbb{R}^2 \cong \mathbb{C}$.

(a) Show that the formula

$$gz = \frac{az + c}{bz + d} \qquad \text{for } z \in \mathfrak{H} \text{ and } g = \begin{bmatrix} a & b \\ c & d \end{bmatrix} \in SL(2, \mathbb{R})$$

defines an action of $SL(2, \mathbb{R})$ by isometries on $\mathfrak{H}$.

(b) Show that this action is transitive on $\mathfrak{H}$.

(c) Show that the stabilizer $\text{Stab}_{SL(2,\mathbb{R})}(i)$ of the point i is

$$SO(2) = \left\{ \begin{bmatrix} \cos\theta, & \sin\theta \\ -\sin\theta & \cos\theta \end{bmatrix} \,\middle|\, \theta \in \mathbb{R} \right\}.$$

(d) The **unit tangent bundle** $T^1\mathfrak{H}$ consists of the tangent vectors of length 1. By differentiation, we obtain an action of $SL(2, \mathbb{R})$ on $T^1\mathfrak{H}$. Show that this action is transitive.

(e) For any unit tangent vector $v \in T^1\mathfrak{H}$, show

$$\text{Stab}_{SL(2,\mathbb{R})}(v) = \pm I.$$

Thus, we may identify $T^1\mathfrak{H}$ with $SL(2, \mathbb{R})/\{\pm I\}$.

(f) It is well known that the geodesics in $\mathfrak{H}$ are semicircles (or lines) that are orthogonal to the real axis. Any $v \in T^1\mathfrak{H}$ is tangent to a unique geodesic. The **geodesic flow** $\hat{y}_t$ on $T^1\mathfrak{H}$ moves the unit tangent vector v

FIGURE 1.1B. The geodesic flow on $\mathfrak{H}$.

a distance t along the geodesic it determines. Show, for some vector v (tangent to the imaginary axis), that, under the identification of Exer. 10e, the geodesic flow $\hat{y}_t$ corresponds to the flow $x \mapsto xa^t$ on $SL(2, \mathbb{R})/\{\pm I\}$, for some $c \in \mathbb{R}$.

(g) The ***horocycles*** in $\mathfrak{H}$ are the circles that are tangent to the real axis (and the lines that are parallel to the real axis). Each $v \in T^1\mathfrak{H}$ is an inward unit normal vector to a unique horocycle H_v. The ***horocycle flow*** $\hat{\eta}_t$ on $T^1\mathfrak{H}$ moves the unit tangent vector v a distance t (counterclockwise, if t is positive) along the corresponding horocycle H_v. Show, for the identification in Exer. 10f, that the horocycle flow corresponds to the flow $x \mapsto xu^t$ on $SL(2, \mathbb{R})/\{\pm I\}$.

#11. Let X be any compact, connected surface of (constant) negative curvature -1. We use the notation and terminology of Exer. 10. It is known that there is a covering map $\rho \colon \mathfrak{H} \to X$ that is a local isometry. Let

$$\Gamma = \{\, y \in SL(2, \mathbb{R}) \mid \rho(yz) = \rho(z) \text{ for all } z \in \mathfrak{H} \,\}.$$

(a) Show that
   (i) Γ is discrete, and
   (ii) $\Gamma \backslash G$ is compact.

(b) Show that the unit tangent bundle T^1X can be identified with $\Gamma \backslash G$, in such a way that
   (i) the geodesic flow on T^1X corresponds to the flow y_t on $\Gamma \backslash SL(2, \mathbb{R})$, and
   (ii) the horocycle flow on T^1X corresponds to the flow η_t on $\Gamma \backslash SL(2, \mathbb{R})$.

#12. Suppose Γ and H are subgroups of a group G. For $x \in G$, let

$$\mathrm{Stab}_H(\Gamma x) = \{\, h \in H \mid \Gamma x h = \Gamma x \,\}$$

be the ***stabilizer*** of Γx in H. Show $\mathrm{Stab}_H(\Gamma x) = x^{-1}\Gamma x \cap H$.

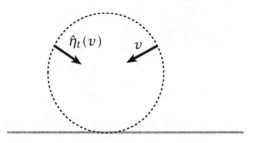

FIGURE 1.1C. The horocycle flow on $\mathfrak{H}$.

#13. Let
  - $G = \mathrm{SL}(2, \mathbb{R})$
  - S be a connected subgroup of G containing $\{u^t\}$, and
  - Γ be a discrete subgroup of G, such that $\Gamma \backslash G$ is compact.

It is known (and you may assume) that

  (a) if $\dim S = 2$, then S is conjugate to the lower-triangular group B,

  (b) if there is a discrete subgroup Λ of S, such that $\Lambda \backslash S$ is compact, then S is **unimodular**, that is, the determinant of the linear transformation $\mathrm{Ad}_S\, g$ is 1, for each $g \in S$, and

  (c) I is the only unipotent matrix in Γ.

Show that if there is a discrete subgroup Λ of S, such that
  - $\Lambda \backslash S$ is compact, and
  - Λ is conjugate to a subgroup of Γ,

then $S = G$.
  [*Hint:* If $\dim S \in \{1, 2\}$, obtain a contradiction.]

#14. Show that if Γ is a discrete subgroup of G, then all fundamental domains for Γ have the same measure. In particular, if one fundamental domain has finite measure, then all do.
  [*Hint:* $\mu(\gamma A) = \mu(A)$, for all $\gamma \in \Gamma$, and every subset A of $\mathcal{F}$.]

#15. Show that if G is **unimodular** (that is, if the left Haar measure is also invariant under right translations) and Γ is a lattice in G, then there is a G-invariant probability measure on $\Gamma \backslash G$.
  [*Hint:* For $A \subset \Gamma \backslash G$, define $\mu_G(A) = \mu(\{\, g \in \mathcal{F} \mid \Gamma g \in A \,\})$.]

#16. Show that if Γ is a lattice in G, then there is a G-invariant probability measure on $\Gamma \backslash G$.

[*Hint:* Use the uniqueness of Haar measure to show, for μ_G as in Exer. 15 and $g \in G$, that there exists $\Delta(g) \in \mathbb{R}^+$, such that $\mu_G(Ag) = \Delta(g)\,\mu_G(A)$ for all $A \subset \Gamma \backslash G$. Then show $\Delta(g) = 1$.]

#17. Show that if Γ is a lattice in G, then the G-invariant probability measure μ_G on $\Gamma \backslash G$ is unique.

[*Hint:* Use μ_G to define a G-invariant measure on G, and use the uniqueness of Haar measure.]

#18. Let
  - $G = \mathrm{SL}(2, \mathbb{R})$,
  - $\Gamma = \mathrm{SL}(2, \mathbb{Z})$,
  - $\mathcal{F} = \{ z \in \mathfrak{H} \mid |z| \geq 1 \text{ and } -1/2 \leq \mathrm{Re}\, z \leq 1/2 \}$, and
  - $e_1 = (1, 0)$ and $e_2 = (0, 1)$,

and define
  - $B\colon G \to \mathbb{R}^2$ by $B(g) = (g^\mathsf{T} e_1, g^\mathsf{T} e_2)$, where g^T denotes the transpose of g,
  - $C\colon \mathbb{R}^2 \to \mathbb{C}$ by $C(x, y) = x + iy$, and
  - $\zeta\colon G \to \mathbb{C}$ by

$$\zeta(g) = \frac{C(g^\mathsf{T} e_2)}{C(g^\mathsf{T} e_1)}.$$

Show:

(a) $\zeta(G) = \mathfrak{H}$,

(b) ζ induces a homeomorphism $\overline{\zeta}\colon \mathfrak{H} \to \mathfrak{H}$, defined by $\overline{\zeta}(gi) = \zeta(g)$,

(c) $\zeta(\gamma g) = \gamma\,\zeta(g)$, for all $g \in G$ and $\gamma \in \Gamma$,

(d) for $g, h \in G$, there exists $\gamma \in \Gamma$, such that $\gamma g = h$ if and only if $\langle g^\mathsf{T} e_1, g^\mathsf{T} e_2 \rangle_{\mathbb{Z}} = \langle h^\mathsf{T} e_1, h^\mathsf{T} e_2 \rangle_{\mathbb{Z}}$, where $\langle v_1, v_2 \rangle_{\mathbb{Z}}$ denotes the abelian group consisting of all integral linear combinations of v_1 and v_2,

(e) for $g \in G$, there exist $v_1, v_2 \in \langle g^\mathsf{T} e_1, g^\mathsf{T} e_2 \rangle_{\mathbb{Z}}$, such that
   (i) $\langle v_1, v_2 \rangle_{\mathbb{Z}} = \langle g^\mathsf{T} e_1, g^\mathsf{T} e_2 \rangle_{\mathbb{Z}}$, and
   (ii) $C(v_2) C(v_1) \in \mathcal{F}$,

(f) $\Gamma \mathcal{F} = \mathfrak{H}$,

(g) if $\gamma \in \Gamma \smallsetminus \{\pm I\}$, then $\gamma \mathcal{F} \cap \mathcal{F}$ has measure 0, and

(h) $\{ g \in G \mid gi \in \mathcal{F} \}$ is is a fundamental domain for Γ in G.

[*Hint:* Choose v_1 and v_2 to be a nonzero vectors of minimal length in $\langle g^\mathsf{T} e_1, g^\mathsf{T} e_2 \rangle_{\mathbb{Z}}$ and $\langle g^\mathsf{T} e_1, g^\mathsf{T} e_2 \rangle_{\mathbb{Z}} \smallsetminus \mathbb{Z} v_1$, respectively.]

#19. Show:

(a) the area element on the hyperbolic plane $\mathfrak{H}$ is $dA = y^{-2}\,dx\,dy$, and

(b) the fundamental domain $\mathcal{F}$ in Fig. 1.1A has finite hyperbolic area.

[*Hint:* We have $\int_a^\infty \int_b^c y^{-2}\,dx\,dy < \infty$.]

#20. Show that if

- Γ is a discrete subgroup of a Lie group G,

- F is a measurable subset of G,

- $\Gamma F = G$, and

- $\mu(F) < \infty$,

then Γ is a lattice in G.

#21. Show that if

- Γ is a discrete subgroup of a Lie group G, and

- $\Gamma \backslash G$ is compact,

then Γ is a lattice in G.

[*Hint:* Show there is a compact subset C of G, such that $\Gamma C = G$, and use Exer. 20.]

#22. Suppose

- Γ is a discrete subgroup of a Lie group G, and

- S is a closed subgroup of G.

Show that if the image $[xS]$ of xS in $\Gamma \backslash G$ is closed, and has finite S-invariant volume, then $(x^{-1}\Gamma x) \cap S$ is a lattice in S.

#23. Let

- Γ be a lattice in a Lie group G,

- $\{x_n\}$ be a sequence of elements of G.

Show that $[x_n]$ has no subsequence that converges in $\Gamma \backslash G$ if and only if there is a sequence $\{y_n\}$ of nonidentity elements of Γ, such that $x_n^{-1}y_n x_n \to e$ as $n \to \infty$.

[*Hint:* ($\Leftarrow$) Contrapositive. If $\{x_{n_k}\} \subset \Gamma C$, where C is compact, then $x_n^{-1}y_n x_n$ is bounded away from e. ($\Rightarrow$) Let $\mathcal{O}$ be a small open subset of G. By passing to a subsequence, we may assume $[x_m\mathcal{O}] \cap [x_n\mathcal{O}] = \varnothing$, for $m \neq n$. Since $\mu(\Gamma \backslash G) < \infty$, then $\mu([x_n\mathcal{O}]) \neq \mu(\mathcal{O})$, for some n. So the natural map $x_n\mathcal{O} \to [x_n\mathcal{O}]$ is not injective. Hence, $x_n^{-1}yx_n \in \mathcal{O}\mathcal{O}^{-1}$ for some $y \in \Gamma$.]

#24. Prove the converse of Exer. 22. That is, if $(x^{-1}\Gamma x) \cap S$ is a lattice in S, then the image $[xS]$ of xS in $\Gamma \backslash G$ is closed (and has finite S-invariant volume).

[*Hint:* Exer. 23 shows that the inclusion of $((x^{-1}\Gamma x) \cap S) \backslash S$ into $\Gamma \backslash G$ is a proper map.]

#25. Let C be the image of the embedding (1.1.18). Assuming that C is closed, show that there is an orbit of the u^t-flow on $\mathrm{SL}(3,\mathbb{Z})\backslash\mathrm{SL}(3,\mathbb{R})$ whose closure is C.

#26. [*Requires some familiarity with hyperbolic geometry*] Let M be a compact, hyperbolic n-manifold, so $M = \Gamma\backslash\mathfrak{H}^n$, for some **discrete** group Γ of isometries of hyperbolic n-space $\mathfrak{H}^n$. For any $k \le n$, there is a natural embedding $\mathfrak{H}^k \hookrightarrow \mathfrak{H}^n$. Composing this with the covering map to M yields a C^∞ immersion $f\colon \mathfrak{H}^k \to M$. Show that if $k \ne 1$, then there is a compact manifold N and a C^∞ function $\psi\colon N \to M$, such that the closure $\overline{f(\mathfrak{H}^k)}$ is equal to $\psi(N)$.

#27. Let $\Gamma = \mathrm{SL}(2,\mathbb{Z})$ and $G = \mathrm{SL}(2,\mathbb{R})$. Use Ratner's Orbit Closure Theorem (and Rem. 1.1.19) to show, for each $g \in G$, that $\Gamma g\Gamma$ is either dense in G or discrete.
[*Hint:* You may assume, without proof, the fact that if N is any connected subgroup of G that is normalized by Γ, then either N is trivial, or $N = G$. (This follows from the Borel Density Theorem (4.7.1).]

1.2. Margulis, Oppenheim, and quadratic forms

Ratner's Theorems have important applications in number theory. In particular, the following result was a major motivating factor. It is often called the "Oppenheim Conjecture," but that terminology is no longer appropriate, because it was proved more than 15 years ago, by G. A. Margulis. See §1.4 for other (more recent) applications.

(1.2.1) **Definition.**

- A (real) **quadratic form** is a homogeneous polynomial of degree 2 (with real coefficients), in any number of variables. For example,

$$Q(x,y,z,w) = x^2 - 2xy + \sqrt{3}yz - 4w^2$$

is a quadratic form (in 4 variables).

- A quadratic form Q is **indefinite** if Q takes both positive and negative values. For example, $x^2 - 3xy + y^2$ is indefinite, but $x^2 - 2xy + y^2$ is definite (see Exer. 2).

- A quadratic form Q in n variables is **nondegenerate** if there does **not** exist a *nonzero* vector $x \in \mathbb{R}^n$, such that $Q(v + x) = Q(v - x)$, for all $v \in \mathbb{R}^n$ (cf. Exer. 3).

(1.2.2) **Theorem** (Margulis). *Let Q be a real, indefinite, non-degenerate quadratic form in $n \geq 3$ variables.*

If Q is not a scalar multiple of a form with integer coefficients, then $Q(\mathbb{Z}^n)$ is dense in $\mathbb{R}$.

(1.2.3) **Example.** If $Q(x, y, z) = x^2 - \sqrt{2}xy + \sqrt{3}z^2$, then Q is not a scalar multiple of a form with integer coefficients (see Exer. 4), so Margulis' Theorem tells us that $Q(\mathbb{Z}^3)$ is dense in $\mathbb{R}$. That is, for each $r \in \mathbb{R}$ and $\epsilon > 0$, there exist $a, b, c \in \mathbb{Z}$, such that $|Q(a, b, c) - r| < \epsilon$.

(1.2.4) **Remark.**

1) The hypothesis that Q is indefinite is necessary. If, say, Q is positive definite, then $Q(\mathbb{Z}^n) \subset \mathbb{R}^{\geq 0}$ is **not** dense in all of $\mathbb{R}$. In fact, if Q is definite, then $Q(\mathbb{Z}^n)$ is **discrete** (see Exer. 7).

2) There are counterexamples when Q has only two variables (see Exer. 8), so the assumption that there are at least 3 variables cannot be omitted in general.

3) A quadratic form is degenerate if (and only if) a change of basis turns it into a form with less variables. Thus, the counterexamples of (2) show the assumption that Q is non-degenerate cannot be omitted in general (see Exer. 9).

4) The converse of Thm. 1.2.2 is true: if $Q(\mathbb{Z}^n)$ is dense in $\mathbb{R}$, then Q cannot be a scalar multiple of a form with integer coefficients (see Exer. 10).

Margulis' Theorem (1.2.2) can be related to Ratner's Theorem by considering the orthogonal group of the quadratic form Q.

(1.2.5) **Definition.**

1) If Q is a quadratic form in n variables, then $\mathrm{SO}(Q)$ is the *orthogonal group* (or *isometry group*) of Q. That is,

$$\mathrm{SO}(Q) = \{\, h \in \mathrm{SL}(n, \mathbb{R}) \mid Q(vh) = Q(v) \text{ for all } v \in \mathbb{R}^n \,\}.$$

(Actually, this is the *special* orthogonal group, because we are including only the matrices of determinant one.)

2) As a special case, $\mathrm{SO}(m, n)$ is a shorthand for the orthogonal group $\mathrm{SO}(Q_{m,n})$, where

$$Q_{m,n}(x_1, \ldots, x_{m+n}) = x_1^2 + \cdots + x_m^2 - x_{m+1}^2 - \cdots - x_{m+n}^2.$$

3) Furthermore, we use $\mathrm{SO}(m)$ to denote $\mathrm{SO}(m, 0)$ (which is equal to $\mathrm{SO}(0, m)$).

(1.2.6) **Definition.** We use $H°$ to denote the *identity component* of a subgroup H of $SL(\ell, \mathbb{R})$; that is, $H°$ is the connected component of H that contains the identity element e. It is a closed subgroup of H.

Because $SO(Q)$ is a real algebraic group (see 4.1.2(8)), Whitney's Theorem (4.1.3) implies that it has only finitely many components. (In fact, it has only one or two components (see Exers. 11 and 13).) Therefore, the difference between $SO(Q)$ and $SO(Q)°$ is very minor, so it may be ignored on a first reading.

Proof of Margulis' Theorem on values of quadratic forms. Let

- $G = SL(3, \mathbb{R})$,

- $\Gamma = SL(3, \mathbb{Z})$,

- $Q_0(x_1, x_2, x_3) = x_1^2 + x_2^2 - x_3^2$, and

- $H = SO(Q_0)° = SO(2, 1)°$.

Let us assume Q has exactly three variables (this causes no loss of generality — see Exer. 15). Then, because Q is indefinite, the signature of Q is either $(2, 1)$ or $(1, 2)$ (cf. Exer. 6); hence, after a change of coordinates, Q must be a scalar multiple of Q_0; thus, there exist $g \in SL(3, \mathbb{R})$ and $\lambda \in \mathbb{R}^\times$, such that

$$Q = \lambda Q_0 \circ g.$$

Note that $SO(Q)° = gHg^{-1}$ (see Exer. 14). Because $H \approx SL(2, \mathbb{R})$ is generated by unipotent elements (see Exer. 16) and $SL(3, \mathbb{Z})$ is a lattice in $SL(3, \mathbb{R})$ (see 4.8.5), we can apply Ratner's Orbit Closure Theorem (see 1.1.19). The conclusion is that there is a connected subgroup S of G, such that

- $H \subset S$,

- the closure of $[gH]$ is equal to $[gS]$, and

- there is an S-invariant probability measure on $[gS]$.

Algebraic calculations show that the only closed, connected subgroups of G that contain H are the two obvious subgroups: G and H (see Exer. 17). Therefore, S must be either G or H. We consider each of these possibilities separately.

Case 1. Assume $S = G$. This implies that

$$\Gamma gH \text{ is dense in } G. \tag{1.2.7}$$

We have

$$
\begin{aligned}
Q(\mathbb{Z}^3) &= Q_0(\mathbb{Z}^3 g) & \text{(definition of } g) \\
&= Q_0(\mathbb{Z}^3 \Gamma g) & (\mathbb{Z}^3 \Gamma = \mathbb{Z}^3) \\
&= Q_0(\mathbb{Z}^3 \Gamma g H) & \text{(definition of } H) \\
&\simeq Q_0(\mathbb{Z}^3 G) & ((1.2.7) \text{ and } Q_0 \text{ is continuous}) \\
&= Q_0(\mathbb{R}^3 \smallsetminus \{0\}) & (vG = \mathbb{R}^3 \smallsetminus \{0\} \text{ for } v \neq 0) \\
&= \mathbb{R},
\end{aligned}
$$

where "$\simeq$" means "is dense in."

Case 2. Assume $S = H$. This is a degenerate case; we will show that Q is a scalar multiple of a form with integer coefficients. To keep the proof short, we will apply some of the theory of algebraic groups. The interested reader may consult Chapter 4 to fill in the gaps.

Let $\Gamma_g = \Gamma \cap (gHg^{-1})$. Because the orbit $[gH] = [gS]$ has finite H-invariant measure, we know that Γ_g is a lattice in $gHg^{-1} = \mathrm{SO}(Q)°$. So the Borel Density Theorem (4.7.1) implies $\mathrm{SO}(Q)°$ is contained in the Zariski closure of Γ_g. Because $\Gamma_g \subset \Gamma = \mathrm{SL}(3, \mathbb{Z})$, this implies that the (almost) algebraic group $\mathrm{SO}(Q)°$ is defined over $\mathbb{Q}$ (see Exer. 4.8#1). Therefore, up to a scalar multiple, Q has integer coefficients (see Exer. 4.8#5). $\quad\square$

Exercises for §1.2.

#1. Suppose α and β are nonzero real numbers, such that α/β is irrational, and define $L(x, y) = \alpha x + \beta y$. Show $L(\mathbb{Z}^2)$ is dense in $\mathbb{R}$. (Margulis' Theorem (1.2.2) is a generalization to quadratic forms of this rather trivial observation about linear forms.)

#2. Let $Q_1(x, y) = x^2 - 3xy + y^2$ and $Q_2(x, y) = x^2 - 2xy + y^2$. Show

(a) $Q_1(\mathbb{R}^2)$ contains both positive and negative numbers, but

(b) $Q_2(\mathbb{R}^2)$ does not contain any negative numbers.

#3. Suppose $Q(x_1, \ldots, x_n)$ is a quadratic form, and let $e_n = (0, \ldots, 0, 1)$ be the nth standard basis vector. Show

$$Q(v + e_n) = Q(v - e_n) \text{ for all } v \in \mathbb{R}^n$$

if and only if there is a quadratic form $Q'(x_1, \ldots, x_{n-1})$ in $n - 1$ variables, such that $Q(x_1, \ldots, x_n) = Q'(x_1, \ldots, x_{n-1})$ for all $x_1, \ldots, x_n \in \mathbb{R}$.

#4. Show that the form Q of Eg. 1.2.3 is not a scalar multiple of a form with integer coefficients; that is, there does not exist $k \in \mathbb{R}^{\times}$, such that all the coefficients of kQ are integers.

#5. Suppose Q is a quadratic form in n variables. Define

$$B : \mathbb{R}^n \times \mathbb{R}^n \to \mathbb{R} \text{ by } B(v, w) = \frac{1}{4}(Q(v + w) - Q(v - w)).$$

(a) Show that B is a symmetric bilinear form on $\mathbb{R}^n$. That is, for $v, v_1, v_2, w \in \mathbb{R}^n$ and $\alpha \in \mathbb{R}$, we have:
   (i) $B(v, w) = B(w, v)$
   (ii) $B(v_1 + v_2, w) = B(v_1, w) + B(v_2, w)$, and
   (iii) $B(\alpha v, w) = \alpha B(v, w)$.

(b) For $h \in \mathrm{SL}(n, \mathbb{R})$, show $h \in \mathrm{SO}(Q)$ if and only if $B(vh, wh) = B(v, w)$ for all $v, w \in \mathbb{R}^n$.

(c) We say that the bilinear form B is **nondegenerate** if for every nonzero $v \in \mathbb{R}^n$, there is some nonzero $w \in \mathbb{R}^n$, such that $B(v, w) \neq 0$. Show that Q is nondegenerate if and only if B is nondegenerate.

(d) For $v \in \mathbb{R}^n$, let $v^{\perp} = \{ w \in \mathbb{R}^n \mid B(v, w) = 0 \}$. Show:
   (i) $v^{\perp}$ is a subspace of $\mathbb{R}^n$, and
   (ii) if B is nondegenerate and $v \neq 0$, then $\mathbb{R}^n = \mathbb{R}v \oplus v^{\perp}$.

#6. (a) Show that $Q_{k,n-k}$ is a nondegenerate quadratic form (in n variables).

(b) Show that $Q_{k,n-k}$ is indefinite if and only if $i \notin \{0, n\}$.

(c) A subspace V of $\mathbb{R}^n$ is **totally isotropic** for a quadratic form Q if $Q(v) = 0$ for all $v \in V$. Show that $\min(k, n - k)$ is the maximum dimension of a totally isotropic subspace for $Q_{k,n-k}$.

(d) Let Q be a nondegenerate quadratic form in n variables. Show there exists a unique $k \in \{0, 1, \ldots, n\}$, such that there is an invertible linear transformation T of $\mathbb{R}^n$ with $Q = Q_{k,n-k} \circ T$. We say that the **signature** of Q is $(k, n - k)$.

[*Hint:* (6d) Choose $v \in \mathbb{R}^n$ with $Q(v) \neq 0$. By induction on n, the restriction of Q to $v^{\perp}$ can be transformed to $Q_{k',n-1-k'}$.]

#7. Let Q be a real quadratic form in n variables. Show that if Q is **positive definite** (that is, if $Q(\mathbb{R}^n) \geq 0$), then $Q(\mathbb{Z}^n)$ is a **discrete** subset of $\mathbb{R}$.

#8. Show:

(a) If α is an irrational root of a quadratic polynomial (with integer coefficients), then there exists $\epsilon > 0$, such that

$$|\alpha - (p/q)| > \epsilon/pq,$$

for all $p, q \in \mathbb{Z}$ (with $p, q \neq 0$).
[*Hint:* $k(x - \alpha)(x - \beta)$ has integer coefficients, for some $k \in \mathbb{Z}^+$ and some $\beta \in \mathbb{R} \smallsetminus \{\alpha\}$.]

(b) The quadratic form $Q(x, y) = x^2 - (3 + 2\sqrt{2})y^2$ is real, indefinite, and nondegenerate, and is **not** a scalar multiple of a form with integer coefficients.

(c) $Q(\mathbb{Z}, \mathbb{Z})$ is not dense in $\mathbb{R}$.
[*Hint:* $\sqrt{3 + 2\sqrt{2}} = 1 + \sqrt{2}$ is a root of a quadratic polynomial.]

#9. Suppose $Q(x_1, x_2)$ is a real, indefinite quadratic form in two variables, and that $Q(x, y)$ is **not** a scalar multiple of a form with integer coefficients, and define $Q^*(y_1, y_2, y_3) = Q(y_1, y_2 - y_3)$.

(a) Show that Q^* is a real, indefinite quadratic form in two variables, and that Q^* is **not** a scalar multiple of a form with integer coefficients.

(b) Show that if $Q(\mathbb{Z}^2)$ is not dense in $\mathbb{R}$, then $Q^*(\mathbb{Z}^3)$ is not dense in $\mathbb{R}$.

#10. Show that if $Q(x_1, \ldots, x_n)$ is a quadratic form, and $Q(\mathbb{Z}^n)$ is dense in $\mathbb{R}$, then Q is **not** a scalar multiple of a form with integer coefficients.

#11. Show that $\mathrm{SO}(Q)$ is connected if Q is definite.
[*Hint:* Induction on n. There is a natural embedding of $\mathrm{SO}(n-1)$ in $\mathrm{SO}(n)$, such that the vector $e_n = (0, 0, \ldots, 0, 1)$ is fixed by $\mathrm{SO}(n-1)$. For $n \geq 2$, the map $\mathrm{SO}(n-1)g \mapsto e_n g$ is a homeomorphism from $\mathrm{SO}(n-1) \backslash \mathrm{SO}(n)$ onto the $(n-1)$-sphere S^{n-1}.]

#12. (Witt's Theorem) Suppose $v, w \in \mathbb{R}^{m+n}$ with $Q_{m,n}(v) = Q_{m,n}(w) \neq 0$, and assume $m + n \geq 2$. Show there exists $g \in \mathrm{SO}(m, n)$ with $vg = w$.
[*Hint:* There is a linear map $T: v^\perp \to w^\perp$ with $Q_{m,n}(xT) = Q_{m,n}(x)$ for all x (see Exer. 6). (Use the assumption $m + n \geq 2$ to arrange for g to have determinant 1, rather than -1.)]

#13. Show that $\mathrm{SO}(m, n)$ has no more than two components if $m, n \geq 1$. (In fact, although you do not need to prove this, it has exactly two components.)
[*Hint:* Similar to Exer. 11. (Use Exer. 12.) If $m > 1$, then $\{v \in \mathbb{R}^{m+n} \mid Q_{m,n} = 1\}$ is connected. The base case $m = n = 1$ should be done separately.]

#14. In the notation of the proof of Thm. 1.2.2, show $SO(Q)^\circ = gHg^{-1}$.

#15. Suppose Q satisfies the hypotheses of Thm. 1.2.2. Show there exist $v_1, v_2, v_3 \in \mathbb{Z}^n$, such that the quadratic form Q' on $\mathbb{R}^3$, defined by $Q'(x_1, x_2, x_3) = Q(x_1 v_1 + x_2 v_2 + x_3 v_3)$, also satisfies the hypotheses of Thm. 1.2.2.
[*Hint:* Choose any v_1, v_2 such that $Q(v_1)/Q(v_2)$ is negative and irrational. Then choose v_3 generically (so Q' is nondegenerate).]

#16. (*Requires some Lie theory*) Show:

(a) The determinant function det is a quadratic form on $\mathfrak{sl}(2, \mathbb{R})$ of signature $(2, 1)$.

(b) The adjoint representation $\mathrm{Ad}_{SL(2,\mathbb{R})}$ maps $SL(2, \mathbb{R})$ into $SO(\det)$.

(c) $SL(2, \mathbb{R})$ is locally isomorphic to $SO(2, 1)^\circ$.

(d) $SO(2, 1)^\circ$ is generated by unipotent elements.

#17. (*Requires some Lie theory*)

(a) Show that $\mathfrak{so}(2, 1)$ is a maximal subalgebra of the Lie algebra $\mathfrak{sl}(3, \mathbb{R})$. That is, there does not exist a subalgebra $\mathfrak{h}$ with $\mathfrak{so}(2, 1) \subsetneq \mathfrak{h} \subsetneq \mathfrak{sl}(3, \mathbb{R})$.

(b) Conclude that if S is any closed, connected subgroup of $SL(3, \mathbb{R})$ that contains $SO(2, 1)$, then

$$\text{either } S = SO(2, 1) \text{ or } S = SL(3, \mathbb{R}).$$

[*Hint:* $\underline{u} = \begin{bmatrix} 0 & 1 & 1 \\ -1 & 0 & 0 \\ 1 & 0 & 0 \end{bmatrix}$ is a nilpotent element of $\mathfrak{so}(2, 1)$, and the kernel of $\mathrm{ad}_{\mathfrak{sl}(3,\mathbb{R})} \underline{u}$ is only 2-dimensional. Since $\mathfrak{h}$ is a submodule of $\mathfrak{sl}(3, \mathbb{R})$, the conclusion follows (see Exer. 4.9#7b).]

1.3. Measure-theoretic versions of Ratner's Theorem

For unipotent flows, Ratner's Orbit Closure Theorem (1.1.14) states that the closure of each orbit is a nice, geometric subset $[xS]$ of the space $X = \Gamma \backslash G$. This means that the orbit is dense in $[xS]$; in fact, it turns out to be **uniformly distributed** in $[xS]$. Before making a precise statement, let us look at a simple example.

(1.3.1) **Example.** As in Eg. 1.1.1, let φ_t be the flow

$$\varphi_t([x]) = [x + tv]$$

on $\mathbb{T}^n$ defined by a vector $v \in \mathbb{R}^n$. Let μ be the Lebesgue measure on $\mathbb{T}^n$, normalized to be a **probability measure** (so $\mu(\mathbb{T}^n) = 1$).

1) Assume $n = 2$, so we may write $v = (a, b)$. If a/b is irrational, then every orbit of φ_t is dense in $\mathbb{T}^2$ (see Exer. 1.1#3). In fact, every orbit is uniformly distributed in $\mathbb{T}^2$: if B is any nice open subset of $\mathbb{T}^2$ (such as an open ball), then the amount of time that each orbit spends in B is proportional to the area of B. More precisely, for each $x \in \mathbb{T}^2$, and letting λ be the Lebesgue measure on $\mathbb{R}$, we have

$$\frac{\lambda(\{t \in [0, T] \mid \varphi_t(x) \in B\})}{T} \to \mu(B) \qquad \text{as } T \to \infty \qquad (1.3.2)$$

(see Exer. 1).

2) Equivalently, if
   - $v = (a, b)$ with a/b irrational,
   - $x \in \mathbb{T}^2$, and
   - f is any continuous function on $\mathbb{T}^2$,

   then

$$\lim_{T \to \infty} \frac{\int_0^T f(\varphi_t(x))\, dt}{T} \to \int_{\mathbb{T}^2} f\, d\mu \qquad (1.3.3)$$

(see Exer. 2).

3) Suppose now that $n = 3$, and assume $v = (a, b, 0)$, with a/b irrational. Then the orbits of φ_t are **not** dense in $\mathbb{T}^3$, so they are **not** uniformly distributed in $\mathbb{T}^3$ (with respect to the usual Lebesgue measure on $\mathbb{T}^3$). Instead, each orbit is uniformly distributed in some subtorus of $\mathbb{T}^3$: given $x = (x_1, x_2, x_3) \in \mathbb{T}$, let μ_2 be the Haar measure on the horizontal 2-torus $\mathbb{T}^2 \times \{x_3\}$ that contains x. Then

$$\frac{1}{T} \int_0^T f(\varphi_t(x))\, dt \to \int_{\mathbb{T}^2 \times \{x_3\}} f\, d\mu_2 \qquad \text{as } T \to \infty$$

(see Exer. 3).

4) In general, for any n and v, and any $x \in \mathbb{T}^n$, there is a subtorus S of $\mathbb{T}^n$, with Haar measure μ_S, such that

$$\int_0^T f(\varphi_t(x))\, dt \to \int_S f\, d\mu_S$$

as $T \to \infty$ (see Exer. 4).

The above example generalizes, in a natural way, to all unipotent flows:

(1.3.4) **Theorem** (Ratner Equidistribution Theorem). *If*
   - *G is any Lie group,*
   - *Γ is any lattice in G, and*

- φ_t *is any unipotent flow on* $\Gamma \backslash G$,

then each φ_t*-orbit is uniformly distributed in its closure.*

(1.3.5) Remark. Here is a more precise statement of Thm. 1.3.4. For any fixed $x \in G$, Ratner's Theorem (1.1.14) provides a connected, closed subgroup S of G (see 1.1.15), such that

1) $\{u^t\}_{t \in \mathbb{R}} \subset S$,

2) the image $[xS]$ of xS in $\Gamma \backslash G$ is closed, and has finite S-invariant volume, and

3) the φ_t-orbit of $[x]$ is dense in $[xS]$.

Let μ_S be the (unique) S-invariant probability measure on $[xS]$. Then Thm. 1.3.4 asserts, for every continuous function f on $\Gamma \backslash G$ with compact support, that

$$\frac{1}{T} \int_0^T f(\varphi_t(x)) \, dt \to \int_{[xS]} f \, d\mu_S \qquad \text{as } T \to \infty.$$

This theorem yields a classification of the φ_t-invariant probability measures.

(1.3.6) Definition. Let

- X be a metric space,
- φ_t be a continuous flow on X, and
- μ be a measure on X.

We say:

1) μ is φ_t-**invariant** if $\mu(\varphi_t(A)) = \mu(A)$, for every Borel subset A of X, and every $t \in \mathbb{R}$.

2) μ is **ergodic** if μ is φ_t-invariant, and every φ_t-invariant Borel function on X is essentially constant (w.r.t. μ). (A function f is **essentially constant** on X if there is a set E of measure 0, such that f is constant on $X \smallsetminus E$.)

Results of Functional Analysis (such as Choquet's Theorem) imply that every invariant probability measure is a convex combination (or, more generally, a direct integral) of ergodic probability measures (see Exer. 6). (See §3.3 for more discussion of the relationship between arbitrary measures and ergodic measures.) Thus, in order to understand all of the invariant measures, it suffices to classify the ergodic ones. Combining Thm. 1.3.4 with the Pointwise Ergodic Theorem (3.1.3) implies that these ergodic measures are of a nice geometric form (see Exer. 7):

(1.3.7) Corollary (Ratner Measure Classification Theorem). *If*

- G *is any Lie group,*

- Γ *is any lattice in G, and*
- φ_t *is any unipotent flow on* $\Gamma \backslash G$,

then every ergodic φ_t*-invariant probability measure on* $\Gamma \backslash G$ *is homogeneous.*

That is, every ergodic φ_t*-invariant probability measure is of the form* μ_S, *for some x and some subgroup S as in Rem. 1.3.5.*

A logical development (and the historical development) of the material proceeds in the opposite direction: instead of deriving Cor. 1.3.7 from Thm. 1.3.4, the main goal of these lectures is to explain the main ideas in a direct proof of Cor. 1.3.7. Then Thms. 1.1.14 and 1.3.4 can be obtained as corollaries. As an illustrative example of this opposite direction — how knowledge of invariant measures can yield information about closures of orbits — let us prove the following classical fact. (A more complete discussion appears in Sect. 1.9.)

(1.3.8) **Definition.** Let φ_t be a continuous flow on a metric space X.

- φ_t is **minimal** if every orbit is dense in X.
- φ_t is **uniquely ergodic** if there is a **unique** φ_t-invariant probability measure on X.

(1.3.9) **Proposition.** *Suppose*

- *G is any Lie group,*
- Γ *is any lattice in G, such that* $\Gamma \backslash G$ *is* **compact**, *and*
- φ_t *is any unipotent flow on* $\Gamma \backslash G$.

If φ_t *is uniquely ergodic, then* φ_t *is minimal.*

Proof. We prove the contrapositive: assuming that some orbit $\varphi_{\mathbb{R}}(x)$ is not dense in $\Gamma \backslash G$, we will show that the G-invariant measure μ_G is not the only φ_t-invariant probability measure on $\Gamma \backslash G$.

Let Ω be the closure of $\varphi_{\mathbb{R}}(x)$. Then Ω is a compact φ_t-invariant subset of $\Gamma \backslash G$ (see Exer. 8), so there is a φ_t-invariant probability measure μ on $\Gamma \backslash G$ that is supported on Ω (see Exer. 9). Because

$$\operatorname{supp} \mu \subset \Omega \subsetneq \Gamma \backslash G = \operatorname{supp} \mu_G,$$

we know that $\mu \neq \mu_G$. Hence, there are (at least) two different φ_t-invariant probability measures on $\Gamma \backslash G$, so φ_t is not uniquely ergodic. $\qquad\square$

(1.3.10) Remark.

1) There is no need to assume Γ is a lattice in Cor. 1.3.7 — the conclusion remains true when Γ is any closed subgroup of G. However, to avoid confusion, let us point out that this is *not* true of the Orbit Closure Theorem — there are counterexamples to (1.1.19) in some cases where $\Gamma \backslash G$ is not assumed to have finite volume. For example, a fractal orbit closure for a^t on $\Gamma \backslash G$ yields a fractal orbit closure for Γ on $\{a^t\} \backslash G$, even though the lattice Γ may be generated by unipotent elements.

2) An appeal to "Ratner's Theorem" in the literature could be referring to any of Ratner's three major theorems: her Orbit Closure Theorem (1.1.14), her Equidistribution Theorem (1.3.4), or her Measure Classification Theorem (1.3.7).

3) There is not universal agreement on the names of these three major theorems of Ratner. For example, the Measure Classification Theorem is also known as "Ratner's Measure-Rigidity Theorem" or "Ratner's Theorem on Invariant Measures," and the Orbit Closure Theorem is also known as the "topological version" of her theorem.

4) Many authors (including M. Ratner) use the adjective *algebraic*, rather than *homogeneous*, to describe measures μ_S as in (1.3.5). This is because μ_S is defined via an algebraic (or, more precisely, group-theoretic) construction.

Exercises for §1.3.

#1. Verify Eg. 1.3.1(1).
[*Hint:* It may be easier to do Exer. 2 first. The characteristic function of B can be approximated by continuous functions.]

#2. Verify Eg. 1.3.1(2); show that if a/b is irrational, and f is any continuous function on $\mathbb{T}^2$, then (1.3.3) holds.
[*Hint:* Linear combinations of functions of the form

$$f(x,y) = \exp 2\pi (mx + ny)i$$

are dense in the space of continuous functions.
Alternate solution: If T_0 is sufficiently large, then, for every $x \in \mathbb{T}^2$, the segment $\{\varphi_t(x)\}_{t=0}^{T_0}$ comes within δ of every point in $\mathbb{T}^2$ (because $\mathbb{T}^2$ is compact and abelian, and the orbits of φ_t are dense). Therefore, the uniform continuity of f implies that if T is sufficiently large, then the value of $(1/T) \int_0^T f(\varphi_t(x)) \, dt$ varies by less than ϵ as x varies over $\mathbb{T}^2$.]

#3. Verify Eg. 1.3.1(3).

#4. Verify Eg. 1.3.1(4).

#5. Let

- φ_t be a continuous flow on a manifold X,
- $\mathrm{Prob}(X)_{\varphi_t}$ be the set of φ_t-invariant Borel probability measures on X, and
- $\mu \in \mathrm{Prob}(X)_{\varphi_t}$.

Show that the following are equivalent:

(a) μ is ergodic;

(b) every φ_t-invariant Borel subset of X is either null or conull;

(c) μ is an ***extreme point*** of $\mathrm{Prob}(X)_{\varphi_t}$, that is, μ is ***not*** a convex combination of two other measures in the space $\mathrm{Prob}(X)_{\varphi_t}$.

[*Hint:* (5a$\Rightarrow$5c) If $\mu = a_1\mu_1 + a_2\mu_2$, consider the Radon-Nikodym derivatives of μ_1 and μ_2 (w.r.t. μ). (5c$\Rightarrow$5b) If A is any subset of X, then μ is the sum of two measures, one supported on A, and the other supported on the complement of A.]

#6. Choquet's Theorem states that if C is any compact subset of a Banach space, then each point in C is of the form $\int_C c \, d\mu(c)$, where ν is a probability measure supported on the extreme points of C. Assuming this fact, show that every φ_t-invariant probability measure is an integral of ergodic φ_t-invariant measures.

#7. Prove Cor. 1.3.7.
[*Hint:* Use (1.3.4) and (3.1.3).]

#8. Let

- φ_t be a continuous flow on a metric space X,
- $x \in X$, and
- $\varphi_{\mathbb{R}}(x) = \{ \varphi_t(x) \mid t \in \mathbb{R} \}$ be the ***orbit*** of x.

Show that the closure $\overline{\varphi_{\mathbb{R}}(x)}$ of $\varphi_{\mathbb{R}}(x)$ is φ_t-***invariant***; that is, $\varphi_t\bigl(\overline{\varphi_{\mathbb{R}}(x)}\bigr) = \overline{\varphi_{\mathbb{R}}(x)}$, for all $t \in \mathbb{R}$.

#9. Let

- φ_t be a continuous flow on a metric space X, and
- Ω be a nonempty, compact, φ_t-invariant subset of X.

Show there is a φ_t-invariant probability measure μ on X, such that $\mathrm{supp}(\mu) \subset \Omega$. (In other words, the complement of Ω is a null set, w.r.t. μ.)

[*Hint:* Fix $x \in \Omega$. For each $n \in \mathbb{Z}^+$, $(1/n) \int_0^n f(\varphi_t(x)) \, dt$ defines a probability measure μ_n on X. The limit of any convergent subsequence is φ_t-invariant.]

#10. Let

- $S^1 = \mathbb{R} \cup \{\infty\}$ be the one-point compactification of $\mathbb{R}$, and

- $\varphi_t(x) = x + t$ for $t \in \mathbb{R}$ and $x \in S^1$.

Show φ_t is a flow on S^1 that is uniquely ergodic (and continuous) but not minimal.

#11. Suppose φ_t is a uniquely ergodic, continuous flow on a compact metric space X. Show φ_t is minimal if and only if there is a φ_t-invariant probability measure μ on X, such that the support of μ is all of X.

#12. Show that the conclusion of Exer. 9 can fail if we omit the hypothesis that Ω is compact.
[*Hint:* Let $\Omega = X = \mathbb{R}$, and define $\varphi_t(x) = x + t$.]

1.4. Some applications of Ratner's Theorems

This section briefly describes a few of the many results that rely crucially on Ratner's Theorems (or the methods behind them). Their proofs require substantial new ideas, so, although we will emphasize the role of Ratner's Theorems, we do not mean to imply that any of these theorems are merely corollaries.

1.4A. Quantitative versions of Margulis' Theorem on values of quadratic forms. As discussed in §1.2, G. A. Margulis proved, under appropriate hypotheses on the quadratic form Q, that the values of Q on $\mathbb{Z}^n$ are dense in $\mathbb{R}$. By a more sophisticated argument, it can be shown (except in some small cases) that the values are uniformly distributed in $\mathbb{R}$, not just dense:

(1.4.1) Theorem. *Suppose*

- *Q is a real, nondegenerate quadratic form,*

- *Q is not a scalar multiple of a form with integer coefficients, and*

- *the signature (p, q) of Q satisfies $p \geq 3$ and $q \geq 1$.*

Then, for any interval (a, b) in $\mathbb{R}$, we have

$$\frac{\#\left\{ v \in \mathbb{Z}^{p+q} \;\middle|\; \begin{array}{c} a < Q(v) < b, \\ \|v\| \leq N \end{array} \right\}}{\mathrm{vol}\left\{ v \in \mathbb{R}^{p+q} \;\middle|\; \begin{array}{c} a < Q(v) < b, \\ \|v\| \leq N \end{array} \right\}} \to 1 \qquad as\ N \to \infty.$$

(1.4.2) Remark.

1) By calculating the appropriate volume, one finds a constant C_Q, depending only on Q, such that, as $N \to \infty$,

$$\#\left\{ v \in \mathbb{Z}^{p+q} \;\middle|\; \begin{array}{c} a < Q(v) < b, \\ \|v\| \leq N \end{array} \right\} \sim (b - a) C_Q N^{p+q-2}.$$

2) The restriction on the signature of Q cannot be eliminated; there are counterexamples of signature $(2, 2)$ and $(2, 1)$.

Why Ratner's Theorem is relevant. We provide only an indication of the direction of attack, not an actual proof of the theorem.

1) Let $K = \mathrm{SO}(p) \times \mathrm{SO}(q)$, so K is a compact subgroup of $\mathrm{SO}(p, q)$.

2) For $c, r \in \mathbb{R}$, it is not difficult to see that K is transitive on
$$\{ v \in \mathbb{R}^{p+q} \mid Q_{p,q}(v) = c, \; \|v\| = r \}$$
(unless $q = 1$, in which case K has two orbits).

3) Fix $g \in \mathrm{SL}(p + q, \mathbb{R})$, such that $Q = Q_{p,q} \circ g$. (Actually, Q may be a scalar multiple of $Q_{p,q} \circ g$, but let us ignore this issue.)

4) Fix a nontrivial one-parameter unipotent subgroup u^t of $\mathrm{SO}(p, q)$.

5) Let $\mathcal{O}$ be a bounded open set that
 • intersects $Q_{p,q}^{-1}(c)$, for all $c \in (a, b)$, and

 • does not contain any fixed points of u^t in its closure.

By being a bit more careful in the choice of $\mathcal{O}$ and u^t, we may arrange that $\|w u^t\|$ is within a constant factor of t^2 for all $w \in \mathcal{O}$ and all large $t \in \mathbb{R}$.

6) If v is any large element of $\mathbb{R}^{p+q}$, with $Q_{p,q}(v) \in (a, b)$, then there is some $w \in \mathcal{O}$, such that $Q_{p,q}(w) = Q_{p,q}(v)$. If we choose $t \in \mathbb{R}^+$ with $\|w u^{-t}\| = \|v\|$ (note that $t < C\sqrt{\|v\|}$, for an appropriate constant C), then $w \in vKu^t$. Therefore
$$\int_0^{C\sqrt{\|v\|}} \int_K \chi_{\mathcal{O}}(vku^t) \, dk \, dt \neq 0, \tag{1.4.3}$$
where $\chi_{\mathcal{O}}$ is the characteristic function of $\mathcal{O}$.

7) We have
$$\left\{ v \in \mathbb{Z}^{p+q} \;\middle|\; \begin{matrix} a < Q(v) < b, \\ \|v\| \leq N \end{matrix} \right\} = \left\{ v \in \mathbb{Z}^{p+q} \;\middle|\; \begin{matrix} a < Q_{p,q}(vg) < b, \\ \|v\| \leq N \end{matrix} \right\}.$$

From (1.4.3), we see that the cardinality of the right-hand side can be approximated by
$$\sum_{v \in \mathbb{Z}^{p+q}} \int_0^{C\sqrt{N}} \int_K \chi_{\mathcal{O}}(vgku^t) \, dk \, dt.$$

8) By
- bringing the sum inside the integrals, and
- defining $\widetilde{\chi_0} \colon \Gamma \backslash G \to \mathbb{R}$ by

$$\widetilde{\chi_0}(\Gamma x) = \sum_{v \in \mathbb{Z}^{p+q}} \chi_0(vx),$$

where $G = \mathrm{SL}(p+q, \mathbb{R})$ and $\Gamma = \mathrm{SL}(p+q, \mathbb{Z})$,
we obtain

$$\int_0^{c\sqrt{N}} \int_K \widetilde{\chi_0}(\Gamma g k u^t) \, dk \, dt. \tag{1.4.4}$$

The outer integral is the type that can be calculated from Ratner's Equidistribution Theorem (1.3.4) (except that the integrand is not continuous and may not have compact support).

(1.4.5) Remark.

1) Because of technical issues, it is actually a more precise version (1.9.5) of equidistribution that is used to estimate the integral (1.4.4). In fact, the issues are so serious that the above argument actually yields only a lower bound on the integral. Obtaining the correct upper bound requires additional difficult arguments.

2) Furthermore, the conclusion of Thm. 1.4.1 fails for some forms of signature $(2, 2)$ or $(2, 1)$; the limit may be $+\infty$.

1.4B. Arithmetic Quantum Unique Ergodicity. Suppose Γ is a lattice in $G = \mathrm{SL}(2, \mathbb{R})$, such that $\Gamma \backslash G$ is compact. Then $M = \Gamma \backslash \mathfrak{H}$ is a compact manifold. (We should assume here that Γ has no elements of finite order.) The hyperbolic metric on $\mathfrak{H}$ yields a Riemannian metric on M, and there is a corresponding Laplacian Δ and volume measure vol (normalized to be a probability measure). Let

$$0 = \lambda_0 < \lambda_1 \le \lambda_2 \le \cdots$$

be the eigenvalues of Δ (with multiplicity). For each λ_n, there is a corresponding eigenfunction ϕ_n, which we assume to be normalized (and real valued), so that $\int_M \phi_n^2 \, d\,\mathrm{vol} = 1$.

In Quantum Mechanics, one may think of ϕ_n as a possible state of a particle in a certain system; if the particle is in this state, then the probability of finding it at any particular location on M is represented by the probability distribution $\phi_n^2 \, d\,\mathrm{vol}$. It is natural to investigate the limit as $\lambda_n \to \infty$, for this describes the behavior that can be expected when there is enough energy that quantum effects can be ignored, and the laws of classical mechanics can be applied.

It is conjectured that, in this classical limit, the particle becomes uniformly distributed:

(1.4.6) Conjecture (Quantum Unique Ergodicity).

$$\lim_{n \to \infty} \phi_n^2 \, d\,\mathrm{vol} = d\,\mathrm{vol}.$$

This conjecture remains open, but it has been proved in an important special case.

(1.4.7) Definition.

1) If Γ belongs to a certain family of lattices (constructed by a certain method from an algebra of quaternions over $\mathbb{Q}$) then we say that Γ is a **congruence** lattice. Although these are very special lattices, they arise very naturally in many applications in number theory and elsewhere.

2) If the eigenvalue λ_n is simple (i.e, if λ_n is not a repeated eigenvalue), then the corresponding eigenfunction ϕ_n is uniquely determined (up to a sign). If λ_n is not simple, then there is an entire space of possibilities for ϕ_n, and this ambiguity results in a serious difficulty.

   Under the assumption that Γ is a congruence lattice, it is possible to define a particular orthonormal basis of each eigenspace; the elements of this basis are well defined (up to a sign) and are called *Hecke eigenfunctions*, (or *Hecke-Maass cusp forms*).

We remark that if Γ is a congruence lattice, and there are no repeated eigenvalues, then each ϕ_n is automatically a Hecke eigenfunction.

(1.4.8) Theorem. *If*

- Γ *is a congruence lattice, and*
- *each ϕ_n is a Hecke eigenfunction,*

then $\lim_{n \to \infty} \phi_n^2 \, d\,\mathrm{vol} = d\,\mathrm{vol}.$

Why Ratner's Theorem is relevant. Let μ be a limit of some subsequence of $\phi_n^2 \, d\,\mathrm{vol}$. Then μ can be lifted to an a^t-invariant probability measure $\hat{\mu}$ on $\Gamma \backslash G$. Unfortunately, a^t is not unipotent, so Ratner's Theorem does not immediately apply.

Because each ϕ_n is assumed to be a Hecke eigenfunction, one is able to further lift μ to a measure $\tilde{\mu}$ on a certain homogeneous space $\tilde{\Gamma} \backslash (G \times \mathrm{SL}(2, \mathbb{Q}_p))$, where $\mathbb{Q}_p$ denotes the field of p-adic numbers for an appropriate prime p. There is an additional action coming from the factor $\mathrm{SL}(2, \mathbb{Q}_p)$. By combining this action with the "Shearing Property" of the u^t-flow, much as in the proof of (1.6.10) below, one shows that $\tilde{\mu}$ is u^t-invariant. (This argument requires one to know that the entropy $h_{\tilde{\mu}}(a^t)$ is nonzero.) Then a version of Ratner's Theorem generalized to apply to p-adic groups implies that $\tilde{\mu}$ is SL(2, ℝ)-invariant.

1.4C. Subgroups generated by lattices in opposite horospherical subgroups.

(1.4.9) **Notation.** For $1 \le k < \ell$, let

- $\mathbb{U}_{k,\ell} = \left\{ g \in \mathrm{SL}(\ell, \mathbb{R}) \mid g_{i,j} = \delta_{i,j} \text{ if } i > k \text{ or } j \le k \right\}$, and

- $\mathbb{V}_{k,\ell} = \left\{ g \in \mathrm{SL}(\ell, \mathbb{R}) \mid g_{i,j} = \delta_{i,j} \text{ if } j > k \text{ or } i \le k \right\}$.

(We remark that $\mathbb{V}_\ell$ is the transpose of $\mathbb{U}_\ell$.)

(1.4.10) **Example.**

$$\mathbb{U}_{3,5} = \left\{ \begin{bmatrix} 1 & 0 & 0 & * & * \\ 0 & 1 & 0 & * & * \\ 0 & 0 & 1 & * & * \\ 0 & 0 & 0 & 1 & 0 \\ 0 & 0 & 0 & 0 & 1 \end{bmatrix} \right\} \text{ and } \mathbb{V}_{3,5} = \left\{ \begin{bmatrix} 1 & 0 & 0 & 0 & 0 \\ 0 & 1 & 0 & 0 & 0 \\ 0 & 0 & 1 & 0 & 0 \\ * & * & * & 1 & 0 \\ * & * & * & 0 & 1 \end{bmatrix} \right\}.$$

(1.4.11) **Theorem.** *Suppose*

- Γ_U *is a lattice in* $\mathbb{U}_{k,\ell}$, *and*
- Γ_V *is a lattice in* $\mathbb{V}_{k,\ell}$,
- *the subgroup* $\Gamma = \langle \Gamma_U, \Gamma_V \rangle$ *is **discrete**, and*
- $\ell \ge 4$.

Then Γ is a lattice in SL$(\ell, \mathbb{R})$.

Why Ratner's Theorem is relevant. Let $\mathcal{U}_{k,\ell}$ be the space of lattices in $\mathbb{U}_{k,\ell}$ and $\mathcal{V}_{k,\ell}$ be the space of lattices in $\mathbb{V}_{k,\ell}$. (Actually, we consider only lattices with the same "covolume" as Γ_U or Γ_V, respectively.) The block-diagonal subgroup SL$(k, \mathbb{R}) \times \mathrm{SL}(\ell - k, \mathbb{R})$ normalizes $\mathbb{U}_{k,\ell}$ and $\mathbb{V}_{k,\ell}$, so it acts by conjugation on $\mathcal{U}_{k,\ell} \times \mathcal{V}_{k,\ell}$. There is a natural identification of this with an action by translations on a homogeneous space of SL$(k\ell, \mathbb{R}) \times \mathrm{SL}(k\ell, \mathbb{R})$, so Ratner's Theorem implies that the closure of the orbit of (Γ_U, Γ_V) is homogeneous (see 1.1.19). This means that there are very few possibilities for the closure. By combining this conclusion with the discreteness of Γ (and other ideas), one can establish that the orbit itself is closed. This implies a certain compatibility between Γ_U and Γ_V, which leads to the desired conclusion.

(1.4.12) **Remark.** For simplicity, we have stated only a very special case of the above theorem. More generally, one can replace

$SL(\ell, \mathbb{R})$ with another simple Lie group of real rank at least 2, and replace $\mathbb{U}_{k,\ell}$ and $\mathbb{V}_{k,\ell}$ with a pair of opposite horospherical subgroups. The conclusion should be that Γ is a lattice in G, but this has only been proved under certain additional technical assumptions.

1.4D. Other results. For the interested reader, we list some of the many additional publications that put Ratner's Theorems to good use in a variety of ways.

- S. Adams: Containment does not imply Borel reducibility, in: S. Thomas, ed., *Set theory (Piscataway, NJ, 1999)*, pages 1–23. Amer. Math. Soc., Providence, RI, 2002. MR 2003j:03059

- A. Borel and G. Prasad: Values of isotropic quadratic forms at S-integral points, *Compositio Math.* 83 (1992), no. 3, 347–372. MR 93j:11022

- N. Elkies and C. T. McMullen: Gaps in $\sqrt{n}$ mod 1 and ergodic theory, *Duke Math. J.* 123 (2004), no. 1, 95–139. MR 2060024

- A. Eskin, H. Masur, and M. Schmoll: Billiards in rectangles with barriers, *Duke Math. J.* 118 (2003), no. 3, 427–463. MR 2004c:37059

- A. Eskin, S. Mozes, and N. Shah: Unipotent flows and counting lattice points on homogeneous varieties, *Ann. of Math.* 143 (1996), no. 2, 253–299. MR 97d:22012

- A. Gorodnik: Uniform distribution of orbits of lattices on spaces of frames, *Duke Math. J.* 122 (2004), no. 3, 549–589. MR 2057018

- J. Marklof: Pair correlation densities of inhomogeneous quadratic forms, *Ann. of Math.* 158 (2003), no. 2, 419–471. MR 2018926

- T. L. Payne: Closures of totally geodesic immersions into locally symmetric spaces of noncompact type, *Proc. Amer. Math. Soc.* 127 (1999), no. 3, 829–833. MR 99f:53050

- V. Vatsal: Uniform distribution of Heegner points, *Invent. Math.* 148 (2002), no. 1, 1–46. MR 2003j:11070

- R. J. Zimmer: Superrigidity, Ratner's theorem, and fundamental groups, *Israel J. Math.* 74 (1991), no. 2-3, 199–207. MR 93b:22019

1.5. Polynomial divergence and shearing

In this section, we illustrate some basic ideas that are used in Ratner's proof that ergodic measures are homogeneous(1.3.7). This will be done by giving direct proofs of some statements that follow easily from her theorem. Our focus is on the group $SL(2, \mathbb{R})$.

(1.5.1) **Notation.** Throughout this section,

- Γ and Γ' are lattices in $SL(2, \mathbb{R})$,
- u^t is the one-parameter unipotent subgroup of $SL(2, \mathbb{R})$ defined in (1.1.5),
- η_t is the corresponding unipotent flow on $\Gamma \backslash SL(2, \mathbb{R})$, and
- η'_t is the corresponding unipotent flow on $\Gamma' \backslash SL(2, \mathbb{R})$.

Furthermore, to provide an easy source of counterexamples,

- a^t is the one-parameter diagonal subgroup of $SL(2, \mathbb{R})$ defined in (1.1.5),
- γ_t is the corresponding geodesic flow on $\Gamma \backslash SL(2, \mathbb{R})$, and
- γ'_t is the corresponding geodesic flow on $\Gamma' \backslash SL(2, \mathbb{R})$.

For convenience,

- we sometimes write X for $\Gamma \backslash SL(2, \mathbb{R})$, and
- we sometimes write X' for $\Gamma' \backslash SL(2, \mathbb{R})$.

Let us begin by looking at one of Ratner's first major results in the subject of unipotent flows.

(1.5.2) **Example.** Suppose Γ is *conjugate* to Γ'. That is, suppose there exists $g \in SL(2, \mathbb{R})$, such that $\Gamma = g^{-1}\Gamma'g$.

Then η_t is *measurably isomorphic* to η'_t. That is, there is a (measure-preserving) bijection $\psi \colon \Gamma \backslash SL(2, \mathbb{R}) \to \Gamma' \backslash SL(2, \mathbb{R})$, such that $\psi \circ \eta_t = \eta'_t \circ \psi$ (a.e.).

Namely, $\psi(\Gamma x) = \Gamma' g x$ (see Exer. 1). One may note that ψ is continuous (in fact, C^∞), not just measurable.

The example shows that if Γ is conjugate to Γ', then η_t is measurably isomorphic to η'_t. (Furthermore, the isomorphism is obvious, not some complicated measurable function.) Ratner proved the converse. As we will see, this is now an easy consequence of Ratner's Measure Classification Theorem (1.3.7), but it was once an important theorem in its own right.

(1.5.3) **Corollary** (Ratner Rigidity Theorem). *If η_t is measurably isomorphic to η'_t, then Γ is conjugate to Γ'.*

This means that if η_t is measurably isomorphic to η'_t, then it is obvious that the two flows are isomorphic, and an isomorphism can be taken to be a nice, C^∞ map. This is a very special property of unipotent flows; in general, it is difficult to decide whether or not two flows are measurably isomorphic, and measurable isomorphisms are usually not C^∞. For example, it can be shown that y_t is always measurably isomorphic to y'_t (even if Γ is not conjugate to Γ'), but there is usually no C^∞ isomorphism. (For the experts: this is because geodesic flows are Bernoulli.)

(1.5.4) **Remark.**

1) A version of Cor. 1.5.3 remains true with any Lie group G in the place of $\mathrm{SL}(2, \mathbb{R})$, and any (ergodic) unipotent flows.

2) In contrast, the conclusion fails miserably for some subgroups that are not unipotent. For example, choose
   - any $n, n' \geq 2$, and
   - any lattices Γ and Γ' in $G = \mathrm{SL}(n, \mathbb{R})$ and $G' = \mathrm{SL}(n', \mathbb{R})$, respectively.

   By embedding a^t in the top left corner of G and G', we obtain (ergodic) flows φ_t and φ'_t on $\Gamma \backslash G$ and $\Gamma' \backslash G'$, respectively.

   There is obviously no C^∞ isomorphism between φ_t and φ'_t, because the homogeneous spaces $\Gamma \backslash G$ and $\Gamma' \backslash G'$ do not have the same dimension (unless $n = n'$). Even so, it turns out that the two flows are measurably isomorphic (up to a change in speed; that is, after replacing φ_t with φ_{ct} for some $c \in \mathbb{R}^\times$). (For the experts: this is because the flows are Bernoulli.)

Proof of Cor. 1.5.3. Suppose $\psi \colon (\eta_t, X) \to (\eta'_t, X')$ is a measurable isomorphism. Consider the graph of ψ:
$$\mathrm{graph}(\psi) = \{\, (x, \psi(x)) \mid x \in X \,\} \subset X \times X'.$$

Because ψ is measure preserving and equivariant, we see that the measure μ_G on X pushes to an ergodic $\eta_t \times \eta'_t$-invariant measure $\mu_\times$ on $X \times X'$ (see Exer. 3).

- Because $\eta_t \times \eta'_t$ is a unipotent flow (see Exer. 4), Ratner's Measure Classification Theorem (1.3.7) applies, so we conclude that the support of $\mu_\times$ is a single orbit of a subgroup S of $\mathrm{SL}(2, \mathbb{R}) \times \mathrm{SL}(2, \mathbb{R})$.
- On the other hand, $\mathrm{graph}(\psi)$ is the support of $\mu_\times$.

We conclude that the graph of ψ is a single S-orbit (a.e.). This implies that ψ is equal to an ***affine map*** (a.e.); that is, ψ the composition of a group homomorphism and a translation (see Exer. 6).

So ψ is of a purely algebraic nature, not a terrible measurable map, and this implies that Γ is conjugate to Γ' (see Exer. 7).   □

We have seen that Cor. 1.5.3 is a consequence of Ratner's Theorem (1.3.7). It can also be proved directly, but the proof does not help to illustrate the ideas that are the main goal of this section, so we omit it. Instead, let us consider another consequence of Ratner's Theorem.

(1.5.5) Definition. A flow (φ_t, Ω) is a **quotient** (or **factor**) of (η_t, X) if there is a measure-preserving Borel function $\psi \colon X \to \Omega$, such that

$$\psi \circ \eta_t = \varphi_t \circ \psi \text{ (a.e.).} \qquad (1.5.6)$$

For short, we may say ψ is (essentially) **equivariant** if (1.5.6) holds.

The function ψ is **not** assumed to be injective. (Indeed, quotients are most interesting when ψ collapses substantial portions of X to single points in Ω.) On the other hand, ψ must be essentially surjective (see Exer. 8).

(1.5.7) Example.

1) If $\Gamma \subset \Gamma'$, then the horocycle flow (η_t', X') is a quotient of (η_t, X) (see Exer. 9).

2) For $v \in \mathbb{R}^n$ and $v' \in \mathbb{R}^{n'}$, let φ_t and φ_t' be the corresponding flows on $\mathbb{T}^n$ and $\mathbb{T}^{n'}$. If
   - $n' < n$, and
   - $v_i' = v_i$ for $i = 1, \ldots, n'$,

   then (φ_t', X') is a quotient of (φ_t, X) (see Exer. 10).

3) The one-point space $\{*\}$ is a quotient of any flow. This is the **trivial** quotient.

(1.5.8) Remark. Suppose (φ_t, Ω) is a quotient of (η_t, X). Then there is a map $\psi \colon X \to \Omega$ that is essentially equivariant. If we define

$$x \sim y \text{ when } \psi(x) = \psi(y),$$

then $\sim$ is an equivalence relation on X, and we may identify Ω with the quotient space $X/\sim$.

For simplicity, let us assume ψ is completely equivariant (not just a.e.). Then the equivalence relation $\sim$ is η_t-invariant; if $x \sim y$, then $\eta_t(x) \sim \eta_t(y)$. Conversely, if $\equiv$ is an η_t-invariant (measurable) equivalence relation on X, then $X/\equiv$ is a quotient of (φ_t, Ω).

Ratner proved, for $G = \mathrm{SL}(2, \mathbb{R})$, that unipotent flows are closed under taking quotients:

(1.5.9) **Corollary** (Ratner Quotients Theorem). *Each nontrivial quotient of $(\eta_t, \Gamma \backslash \mathrm{SL}(2, \mathbb{R}))$ is isomorphic to a unipotent flow $(\eta'_t, \Gamma' \backslash \mathrm{SL}(2, \mathbb{R}))$, for some lattice Γ'.*

One can derive this from Ratner's Measure Classification Theorem (1.3.7), by putting an $(\eta_t \times \eta_t)$-invariant probability measure on

$$\{ (x, y) \in X \times X \mid \psi(x) = \psi(y) \}.$$

We omit the argument (it is similar to the proof of Cor. 1.5.3 (see Exer. 11)), because it is very instructive to see a direct proof that does not appeal to Ratner's Theorem. However, we will prove only the following weaker statement. (The proof of (1.5.9) can then be completed by applying Cor. 1.8.1 below (see Exer. 1.8#1).)

(1.5.10) **Definition.** A Borel function $\psi \colon X \to \Omega$ has **finite fibers** (a.e.) if there is a conull subset X_0 of X, such that $\psi^{-1}(\omega) \cap X_0$ is finite, for all $\omega \in \Omega$.

(1.5.11) **Example.** If $\Gamma \subset \Gamma'$, then the natural quotient map $\psi \colon X \to X'$ (cf. 1.5.7(1)) has finite fibers (see Exer. 13).

(1.5.12) **Corollary** (Ratner). *If $(\eta_t, \Gamma \backslash \mathrm{SL}(2, \mathbb{R})) \to (\varphi_t, \Omega)$ is any quotient map (and Ω is nontrivial), then ψ has finite fibers (a.e.).*

In preparation for the direct proof of this result, let us develop some basic properties of unipotent flows that are also used in the proofs of Ratner's general theorems.

Recall that

$$u^t = \begin{bmatrix} 1 & 0 \\ t & 1 \end{bmatrix} \quad \text{and} \quad a^t = \begin{bmatrix} e^t & 0 \\ 0 & e^{-t} \end{bmatrix}.$$

For convenience, let $G = \mathrm{SL}(2, \mathbb{R})$.

(1.5.13) **Definition.** If x and y are any two points of $\Gamma \backslash G$, then there exists $q \in G$, such that $y = xq$. If x is close to y (which we denote $x \approx y$), then q may be chosen close to the identity. Thus, we may define a metric d on $\Gamma \backslash G$ by

$$d(x, y) = \min \left\{ \|q - I\| \;\middle|\; \begin{array}{l} q \in G, \\ xq = y \end{array} \right\},$$

where

- I is the identity matrix, and

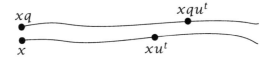

FIGURE 1.5A. The η_t-orbits of two nearby orbits.

- $\| \cdot \|$ is any (fixed) matrix norm on $\mathrm{Mat}_{2 \times 2}(\mathbb{R})$. For example, one may take

$$\left\| \begin{bmatrix} a & b \\ c & d \end{bmatrix} \right\| = \max\{|a|, |b|, |c|, |d|\}.$$

A crucial part of Ratner's method involves looking at what happens to two nearby points as they move under the flow η_t. Thus, we consider two points x and xq, with $q \approx I$, and we wish to calculate $d(\eta_t(x), \eta_t(xq))$, or, in other words,

$$d(xu^t, xqu^t)$$

(see Fig. 1.5A).

- To get from x to xq, one multiplies by q; therefore, $d(x, xq) = \|q - I\|$.
- To get from xu^t to xqu^t, one multiplies by $u^{-t}qu^t$; therefore

$$d(xu^t, xqu^t) = \|u^{-t}qu^t - I\|$$

(as long as this is small — there are infinitely many elements g of G with $xu^t g = xqu^t$, and the distance is obtained by choosing the smallest one, which may not be $u^{-t}qu^t$ if t is large).

Letting

$$q - I = \begin{bmatrix} a & b \\ c & d \end{bmatrix},$$

a simple matrix calculation (see Exer. 14) shows that

$$u^{-t}qu^t - I = \begin{bmatrix} a + bt & b \\ c - (a - d)t - bt^2 & d - bt \end{bmatrix}. \qquad (1.5.14)$$

All the entries of this matrix are polynomials (in t), so we have following obvious conclusion:

(1.5.15) Proposition (Polynomial divergence). *Nearby points of $\Gamma \backslash G$ move apart at polynomial speed.*

In contrast, nearby points of the geodesic flow move apart at exponential speed:

$$a^{-t}qa^t - I = \begin{bmatrix} a & be^{-2t} \\ ce^{2t} & d \end{bmatrix} \qquad (1.5.16)$$

(see Exer. 1.5.16). Intuitively, one should think of polynomial speed as "slow" and exponential speed as "fast." Thus,

- nearby points of a unipotent flow drift slowly apart, but
- nearby points of the geodesic flow jump apart rather suddenly.

More precisely, note that

1) if a polynomial (of bounded degree) stays small for a certain length of time, then it must remain fairly small for a proportional length of time (see Exer. 17):
   - if the polynomial is small for a minute, then it must stay fairly small for another second (say);
   - if the polynomial is small for an hour, then it must stay fairly small for another minute;
   - if the polynomial is small for a year, then it must stay fairly small for another week;
   - if the polynomial is small for several thousand years, then it must stay fairly small for at least a few more decades;
   - if the polynomial has been small for an infinitely long time, then it must stay small forever (in fact, it is constant).

2) In contrast, the exponential function e^t is fairly small ($<$ 1) infinitely far into the past (for $t < 0$), but it becomes arbitrarily large in finite time.

Thus,

1) If two points of a unipotent flow stay close together 90% of the time, then they must stay fairly close together all of the time.

2) In contrast, two points of a geodesic flow may stay close together 90% of the time, but spend the remaining 10% of their lives wandering quite freely (and independently) around the manifold.

The upshot is that if we can get good bounds on a unipotent flow most of the time, then we have bounds that are nearly as good all of the time:

(1.5.17) **Notation.** For convenience, let $x_t = xu^t$ and $y_t = yu^t$.

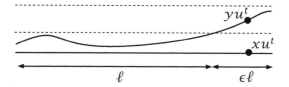

FIGURE 1.5B. Polynomial divergence: Two points that stay close together for a period of time of length ℓ must stay fairly close for an additional length of time $\epsilon\ell$ that is proportional to ℓ.

(1.5.18) Corollary. *For any $\epsilon > 0$, there is a $\delta > 0$, such that if $d(x_t, y_t) < \delta$ for 90% of the times t in an interval $[a, b]$, then $d(x_t, y_t) < \epsilon$ for **all** of the times t in the interval $[a, b]$.*

(1.5.19) Remark. Babysitting provides an analogy that illustrates this difference between unipotent flows and geodesic flows.

1) A unipotent child is easy to watch over. If she sits quietly for an hour, then we may leave the room for a few minutes, knowing that she will not get into trouble while we are away. Before she leaves the room, she will start to make little motions, squirming in her chair. Eventually, as the motions grow, she may get out of the chair, but she will not go far for a while. It is only after giving many warning signs that she will start to walk slowly toward the door.

2) A geodesic child, on the other hand, must be watched almost constantly. We can take our attention away for only a few seconds at a time, because, even if she has been sitting quietly in her chair all morning (or all week), the child might suddenly jump up and run out of the room while we are not looking, getting into all sorts of mischief. Then, before we notice she left, she might go back to her chair, and sit quietly again. We may have no idea there was anything amiss while we were not watching.

Consider the RHS of Eq. 1.5.14, with a, b, c, and d very small. Indeed, let us say they are infinitesimal; too small to see. As t grows, it is the the bottom left corner that will be the first matrix entry to attain macroscopic size (see Exer. 18). Comparing with the definition of u^t (see 1.1.5), we see that this is exactly the direction of the u^t-orbit (see Fig. 1.5C). Thus:

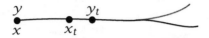

FIGURE 1.5C. Shearing: If two points start out so close together that we cannot tell them apart, then the first difference we see will be that one gets ahead of the other, but (apparently) following the same path. It is only much later that we will detect any difference between their paths.

(1.5.20) **Proposition** (Shearing Property). *The fastest relative motion between two nearby points is parallel to the orbits of the flow.*

The only exception is that if $q \in \{u^t\}$, then $u^{-t}qu^t = q$ for all t; in this case, the points x_t and y_t simply move along together at exactly the same speed, with no relative motion.

(1.5.21) **Corollary.** *If x and y are nearby points, then either*

1) *there exists $t > 0$, such that $y_t \approx x_{t \pm 1}$, or*

2) $y = x_\epsilon$, *for some $\epsilon \approx 0$.*

(1.5.22) **Remark** (Infinitesimals). Many theorems and proofs in these notes are presented in terms of infinitesimals. (We write $x \approx y$ if the distance from x to y is infinitesimal.) There are two main reasons for this:

1) Most importantly, these lectures are intended more to communicate ideas than to record rigorous proofs, and the terminology of infinitesimals is very good at that. It is helpful to begin by pretending that points are infinitely close together, and see what will happen. If desired, the reader may bring in epsilons and deltas after attaining an intuitive understanding of the situation.

2) Nonstandard Analysis is a theory that provides a rigorous foundation to infinitesimals — almost all of the infinitesimal proofs that are sketched here can easily be made rigorous in these terms. For those who are comfortable with it, the infinitesimal approach is often simpler than the classical notation, but we will provide non-infinitesimal versions of the main results in Chap. 5.

(1.5.23) **Remark.** In contrast to the above discussion of u^t,

• the matrix a^t is diagonal, but

• the largest entry in the RHS of Eq. 1.5.16 is an off-diagonal entry,

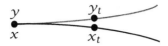

FIGURE 1.5D. Exponential divergence: when two points start out so close together that we cannot tell them apart, the first difference we see may be in a direction transverse to the orbits.

so points in the geodesic flow move apart (at exponential speed) in a direction transverse to the orbits (see Fig. 1.5D).

Let us now illustrate how to use the Shearing Property.

Proof of Cor. 1.5.12. To bring the main ideas to the foreground, let us first consider a special case with some (rather drastic) simplifying assumptions. We will then explain that the assumptions are really not important to the argument.

A1) Let us assume that X is compact (rather than merely having finite volume).

A2) Because (φ_t, Ω) is ergodic (see Exer. 16) and nontrivial, we know that the set of fixed points has measure zero; let us assume that (φ_t, Ω) has no fixed points at all. Therefore,

$$d(\varphi_1(\omega), \omega) \text{ is bounded away from } 0, \atop \text{as } \omega \text{ ranges over } \Omega \qquad (1.5.24)$$

(see Exer. 19).

A3) Let us assume that the quotient map ψ is uniformly continuous (rather than merely being measurable). This may seem unreasonable, but Lusin's Theorem (Exer. 21) tells us that ψ is uniformly continuous on a set of measure $1 - \epsilon$, so, as we shall see, this is actually not a major issue.

Suppose some fiber $\psi^{-1}(\omega_0)$ is infinite. (This will lead to a contradiction.)

Because X is compact, the infinite set $\psi^{-1}(\omega_0)$ must have an accumulation point. Thus, there exist $x \approx y$ with $\psi(x) = \psi(y)$. Because ψ is equivariant, we have

$$\psi(x_t) = \psi(y_t) \text{ for all } t. \qquad (1.5.25)$$

Flow along the orbits until the points x_t and y_t have diverged to a reasonable distance; say, $d(x_t, y_t) = 1$, and let

$$\omega = \psi(y_t). \qquad (1.5.26)$$

Then the Shearing Property implies (see 1.5.21) that

$$y_t \approx \eta_1(x_t). \qquad (1.5.27)$$

Therefore

$$\begin{aligned}
\omega &= \psi(y_t) & (1.5.26) \\
&\approx \psi(\eta_1(x_t)) & ((1.5.27) \text{ and } \psi \text{ is uniformly continuous}) \\
&= \varphi_1(\psi(x_t)) & (\psi \text{ is equivariant}) \\
&= \varphi_1(\omega) & ((1.5.25) \text{ and } (1.5.26)).
\end{aligned}$$

This contradicts (1.5.24).

To complete the proof, we now indicate how to eliminate the assumptions (A1), (A2), and (A3).

First, let us point out that (A1) was not necessary. The proof shows that $\psi^{-1}(\omega)$ has no accumulation point (a.e.); thus, $\psi^{-1}(\omega)$ must be countable. Measure theorists can show that a countable-to-one equivariant map between ergodic spaces with invariant probability measure must actually be finite-to-one (a.e.) (see Exer. 3.3#3). Second, note that it suffices to show, for each $\epsilon > 0$, that there is a subset $\hat{X}$ of X, such that

- $\mu(\hat{X}) > 1 - \epsilon$ and
- $\psi^{-1}(\omega) \cap \hat{X}$ is countable, for a.e. $\omega \in \Omega$.

Now, let $\hat{\Omega}$ be the complement of the set of fixed points in Ω. This is conull, so $\psi^{-1}(\hat{\Omega})$ is conull in X. Thus, by Lusin's Theorem, $\psi^{-1}(\hat{\Omega})$ contains a compact set K, such that

- $\mu_G(K) > 0.99$, and
- ψ is uniformly continuous on K.

Instead of making assumptions (A2) and (A3), we work inside of K. Note that:

(A2′) $d(\varphi_1(\omega), \omega)$ is bounded away from 0, for $\omega \in \psi(K)$; and

(A3′) ψ is uniformly continuous on K.

Let $\hat{X}$ be a generic set for K; that is, points in $\hat{X}$ spend 99% of their lives in K. The Pointwise Ergodic Theorem (3.1.3) tells us that the generic set is conull. (Technically, we need the points of $\hat{X}$ to be ***uniformly generic***: there is a constant L, independent of x, such that

for all $L' > L$ and $x \in \hat{X}$, at least 98% of
the initial segment $\{\phi_t(x)\}_{t=0}^{L'}$ is in K,

and this holds only on a set of measure $1 - \epsilon$, but let us ignore this detail.) Given $x, y \in \hat{X}$, with $x \approx y$, flow along the orbits until $d(x_t, y_t) = 1$. Unfortunately, it may not be the case that x_t

and y_t are in K, but, because 99% of each orbit is in K, we may choose a nearby value t' (say, $t \leq t' \leq 1.1t$), such that

$$x_{t'} \in K \text{ and } y_{t'} \in K.$$

By polynomial divergence, we know that the y-orbit drifts *slowly* ahead of the x-orbit, so

$$y_{t'} \approx \eta_{1+\delta}(x_{t'}) \text{ for some small } \delta.$$

Thus, combining the above argument with a strengthened version of (A2′) (see Exer. 22) shows that $\psi^{-1}(\omega) \cap \hat{X}$ has no accumulation points (hence, is countable). This completes the proof. □

The following application of the Shearing Property is a better illustration of how it is actually used in the proof of Ratner's Theorem.

(1.5.28) Definition. A *self-joining* of (η_t, X) is a probability measure $\hat{\mu}$ on $X \times X$, such that

1) $\hat{\mu}$ is invariant under the diagonal flow $\eta_t \times \eta_t$, and

2) $\hat{\mu}$ projects to μ_G on each factor of the product; that is, $\hat{\mu}(A \times Y) = \mu_G(A)$ and $\hat{\mu}(Y \times B) = \mu_G(B)$.

(1.5.29) Example.

1) The product measure $\hat{\mu} = \mu_G \times \mu_G$ is a self-joining.

2) There is a natural diagonal embedding $x \mapsto (x, x)$ of X in $X \times X$. This is clearly equivariant, so μ_G pushes to an $(\eta_t \times \eta_t)$-invariant measure on $X \times X$. It is a self-joining, called the *diagonal self-joining*.

3) Replacing the identity map $x \mapsto x$ with covering maps yields a generalization of (2): For some $g \in G$, let $\Gamma' = \Gamma \cap (g^{-1}\Gamma g)$, and assume Γ' has finite index in Γ. There are two natural covering maps from X' to X:
   - $\psi_1(\Gamma' x) = \Gamma x$, and
   - $\psi_2(\Gamma' x) = \Gamma g x$

   (see Exer. 23). Define $\psi \colon X' \to X \times X$ by

   $$\psi(x) = (\psi_1(x), \psi_2(x)).$$

   Then
   - ψ is equivariant (because ψ_1 and ψ_2 are equivariant), so the G-invariant measure μ'_G on X' pushes to an invariant measure $\hat{\mu} = \psi_* \mu'_G$ on $X \times X$, defined by

   $$(\psi_* \mu'_G)(A) = \mu'_G(\psi^{-1}(A)),$$

   and

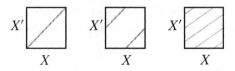

FIGURE 1.5E. The diagonal self-joining and some other finite-cover self-joinings.

- $\hat{\mu}$ is a self-joining (because ψ_1 and ψ_2 are measure preserving).

This is called a **finite-cover self-joining**.

For unipotent flows on $\Gamma \backslash \mathrm{SL}(2, \mathbb{R})$, Ratner showed that these are the only product self-joinings.

(1.5.30) **Corollary** (Ratner's Joinings Theorem). *Any ergodic self-joining of a horocycle flow must be either*

  1) *a finite cover, or*

  2) *the product self-joining.*

This follows quite easily from Ratner's Theorem (1.3.7) (see Exer. 24), but we give a direct proof of the following weaker statement. (Note that if the self-joining $\hat{\mu}$ is a finite cover, then $\hat{\mu}$ has finite fibers; that is, μ is supported on a set with only finitely many points from each horizontal or vertical line (see Exer. 25)). Corollary 1.8.1 will complete the proof of (1.5.30).

(1.5.31) **Corollary.** *If $\hat{\mu}$ is an ergodic self-joining of η_t, then either*

  1) *$\hat{\mu}$ is the product joining, or*

  2) *$\hat{\mu}$ has finite fibers.*

Proof. We omit some details (see Exer. 26 and Rem. 1.5.33).

Consider two points (x, a) and (x, b) in the same vertical fiber. If the fiber is infinite (and X is compact), then we may assume $a \approx b$. By the Shearing Property (1.5.21), there is some t with $a_t \approx \eta_1(b_t)$. Let ξ_t be the **vertical flow** on $X \times X$, defined by

$$\xi_t(x, y) = (x, \eta_t(y)).$$

Then

$$(x, a)_t = (x_t, a_t) \approx (x_t, \eta_1(b_t)) = \xi_1((x, b)_t).$$

We now consider two cases.

Case 1. Assume $\hat{\mu}$ is ξ_t-invariant. Then the ergodicity of η_t implies that $\hat{\mu}$ is the product joining (see Exer. 27).

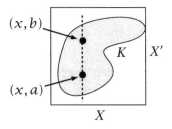

FIGURE 1.5F. Two points (x, a) and (x, b) on the same vertical fiber.

*Case 2. Assume $\hat{\mu}$ is **not** ξ_t-invariant.* In other words, we have $(\xi_1)_*(\hat{\mu}) \neq \hat{\mu}$. On the other hand, $(\xi_1)_*(\hat{\mu})$ is η_t-invariant (because ξ_1 commutes with η_t (see Exer. 28)). It is a general fact that any two ergodic measures for the same flow must be mutually singular (see Exer. 30), so $(\xi_1)_*(\hat{\mu}) \perp \hat{\mu}$; thus, there is a conull subset $\hat{X}$ of $X \times X$, such that $\hat{\mu}(\xi_1(X)) = 0$. From this, it is not difficult to see that there is a compact subset K of $X \times X$, such that

$$\hat{\mu}(K) > 0.99 \text{ and } d(K, \xi_1(K)) > 0 \qquad (1.5.32)$$

(see Exer. 31).

To complete the proof, we show:

Claim. Any generic set for K intersects each vertical fiber $\{x\} \times X$ in a countable set. Suppose not. (This will lead to a contradiction.) Because the fiber is uncountable, there exist (x, a) and (x, b) in the generic set, with $a \approx b$. Flow along the orbits until

$$a_t \approx \eta_1(b_t),$$

and assume (as is true 98% of the time) that $(x, a)_t$ and $(x, b)_t$ belong to K. Then

$$K \ni (x, a)_t = (x_t, a_t) \approx (x_t, \eta_1(b_t)) = \xi_1((x, b)_t) \in \xi_1(K),$$

so $d(K, \xi_1(K)) = 0$. This contradicts (1.5.32). $\qquad\square$

(1.5.33) **Remark.** The above proof ignores an important technical point that also arose on p. 40 of the proof of Cor. 1.5.12: at the precise time t when $a_t \approx \eta_1(b_t)$, it may not be the case that $(x, a)_t$ and $(x, b)_t$ belong to K. We choose a nearby value t' (say, $t \leq t' \leq 1.1t$), such that $(x, a)_{t'}$ and $(x, b)_{t'}$ belong to K. By polynomial divergence, we know that $a_{t'} \approx \eta_{1+\delta}(b_{t'})$ for some small δ.

Hence, the final stage of the proof requires $\xi_{1+\delta}(K)$ to be disjoint from K. Since K must be chosen before we know the value of δ, this is a serious issue. Fortunately, a single set K can be chosen that works for all values of δ (cf. 5.8.6).

Exercises for §1.5.

#1. Suppose Γ and Γ' are lattices in $G = \mathrm{SL}(2, \mathbb{R})$. Show that if $\Gamma' = g^{-1}\Gamma g$, for some $g \in G$, then the map $\psi \colon \Gamma \backslash \mathrm{SL}(2, \mathbb{R}) \to \Gamma' \backslash \mathrm{SL}(2, \mathbb{R})$, defined by $\psi(\Gamma x) = \Gamma' g x$,

　(a) is well defined, and

　(b) is equivariant; that is, $\psi \circ \eta_t = \eta'_t \circ \psi$.

#2. A nonempty, closed subset C of $X \times X$ is **minimal** for $\eta_t \times \eta_t$ if the orbit of every point in C is dense in C. Show that if C is a compact minimal set for $\eta_t \times \eta_t$, then C has finite fibers.

[*Hint:* Use the proof of (1.5.31).]

#3. Suppose
 - (X, μ) and (X', μ') are Borel measure spaces,
 - φ_t and φ'_t are (measurable) flows on X and X', respectively,
 - $\psi \colon X \to X'$ is a measure-preserving map, such that $\psi \circ \varphi_t = \varphi'_t \circ \psi$ (a.e.), and
 - $\mu_\times$ is the Borel measure on $X \times X'$ that is defined by

$$\mu_\times(\Omega) = \mu\{ x \in X \mid (x, \psi(x)) \in \Omega \}.$$

Show:

　(a) $\mu_\times$ is $\varphi_t \times \varphi'_t$-invariant.

　(b) If μ is ergodic (for φ_t), then $\mu_\times$ is ergodic (for $\varphi_t \times \varphi'_t$).

#4. The product $\mathrm{SL}(2, \mathbb{R}) \times \mathrm{SL}(2, \mathbb{R})$ has a natural embedding in $\mathrm{SL}(4, \mathbb{R})$ (as block diagonal matrices). Show that if u and v are unipotent matrices in $\mathrm{SL}(2, \mathbb{R})$, then the image of (u, v) is a unipotent matrix in $\mathrm{SL}(4, \mathbb{R})$.

#5. Suppose ψ is a function from a group G to a group H. Show ψ is a homomorphism if and only if the graph of ψ is a subgroup of $G \times H$.

#6. Suppose
 - G_1 and G_2 are groups,
 - Γ_1 and Γ_2 are subgroups of G_1 and G_2, respectively,
 - ψ is a function from $\Gamma_1 \backslash G_1$ to $\Gamma_2 \backslash G_2$,
 - S is a subgroup of $G_1 \times G_2$, and
 - the graph of ψ is equal to xS, for some $x \in \Gamma_1 \backslash G_1 \times \Gamma_2 \backslash G_2$.

Show:

(a) If $S \cap (e \times G_2)$ is trivial, then ψ is an affine map.

(b) If ψ is surjective, and Γ_2 does not contain any nontrivial normal subgroup of G_2, then $S \cap (e \times G_2)$ is trivial.

[*Hint:* (6a) S is the graph of a homomorphism from G_1 to G_2 (see Exer. 5).]

#7. Suppose Γ and Γ' are lattices in a (simply connected) Lie group G.

(a) Show that if there is a bijective affine map from $\Gamma \backslash G$ to $\Gamma' \backslash G$, then there is an automorphism α of G, such that $\alpha(\Gamma) = \Gamma'$.

(b) Show that if α is an automorphism of $SL(2, \mathbb{R})$, such that $\alpha(u)$ is conjugate to u, then α is an **inner automorphism**; that is, there is some $g \in SL(2, \mathbb{R})$, such that $\alpha(x) = g^{-1}xg$, for all $x \in SL(2, \mathbb{R})$.

(c) Show that if there is a bijective affine map $\psi \colon \Gamma \backslash G \to \Gamma' \backslash G$, such that $\psi(xu) = \psi(x)u$, for all $x \in \Gamma \backslash G$, then Γ is conjugate to Γ'.

#8. Show that if $\psi \colon X \to \Omega$ is a measure-preserving map, then $\psi(X)$ is a conull subset of Ω.

#9. Verify Eg. 1.5.7(1).

#10. Verify Eg. 1.5.7(2).

#11. Give a short proof of Cor. 1.5.9, by using Ratner's Measure Classification Theorem.

#12. Suppose Γ is a lattice in a Lie group G. Show that a subgroup Γ' of Γ is a lattice if and only if Γ' has finite index in Γ.

#13. Suppose Γ and Γ' are lattices in a Lie group G, such that $\Gamma \subset \Gamma'$. Show that the natural map $\Gamma \backslash G \to \Gamma' \backslash G$ has finite fibers.

#14. Verify Eq. (1.5.14).

#15. Verify Eq. (1.5.16).

#16. Suppose (φ'_t, Ω') is a quotient of a flow (φ_t, Ω). Show that if φ is ergodic, then φ' is ergodic.

#17. Given any natural number d, and any $\delta > 0$, show there is some $\epsilon > 0$, such that if

- $f(x)$ is any real polynomial of degree $\leq d$,
- $C \in \mathbb{R}^+$,
- $[k, k + \ell]$ is any real interval, and
- $|f(t)| < C$ for all $t \in [k, k + \ell]$,

then $|f(t)| < (1 + \delta)C$ for all $t \in [k, k + (1 + \epsilon)\ell]$.

#18. Given positive constants $\epsilon < L$, show there exists $\epsilon_0 > 0$, such that if $|\alpha|, |b|, |c|, |d| < \epsilon_0$, and $N > 0$, and we have

$$|c - (\alpha - d)t - bt^2| < L \text{ for all } t \in [0, N],$$

then $|\alpha + bt| + |d - bt| < \epsilon$ for all $t \in [0, N]$.

#19. Suppose ψ is a homeomorphism of a compact metric space (X, d), and that ψ has no fixed points. Show there exists $\epsilon > 0$, such that, for all $x \in X$, we have $d(\psi(x), x) > \epsilon$.

#20. (Probability measures are regular) Suppose
   - X is a metric space that is separable and locally compact,
   - μ is a Borel probability measure on X,
   - $\epsilon > 0$, and
   - A is a measurable subset of X.

Show:

(a) there exist a compact set C and an open set V, such that $C \subset A \subset V$ and $\mu(V \smallsetminus C) < \epsilon$, and

(b) there is a continuous function f on X, such that

$$\mu\{x \in X \mid \chi_A(x) \neq f(x)\} < \epsilon,$$

where χ_A is the characteristic function of A.

[*Hint:* Recall that "**separable**" means X has a countable, dense subset, and that "**locally compact**" means every point of X is contained in an open set with compact closure. (20a) Show the collection $\mathcal{A}$ of sets A such that C and V exist for every ϵ is a σ-algebra. (20b) Note that

$$\frac{d(x, X \smallsetminus V)}{d(x, X \smallsetminus V) + d(x, C)}$$

is a continuous function that is 1 on C and 0 outside of V.]

#21. (Lusin's Theorem) Suppose
   - X is a metric space that is separable and locally compact,
   - μ is a Borel probability measure on X,
   - $\epsilon > 0$, and
   - $\psi \colon X \to \mathbb{R}$ is measurable.

Show there is a continuous function f on X, such that

$$\mu\{x \in X \mid \psi(x) \neq f(x)\} < \epsilon.$$

[*Hint:* Construct step functions ψ_n that converge uniformly to ψ on a set of measure $1 - (\epsilon/2)$. (Recall that a **step function** is a linear combination of characteristic functions of sets.) Now ψ_n is equal to a continuous function f_n on a set of measure $1 - 2^{-n}$

(cf. Exer. 20). Then $\{f_n\}$ converges to f uniformly on a set of measure $> 1 - \epsilon$.]

#22. Suppose
- (X, d) is a metric space,
- μ is a probability measure on X,
- φ_t is an ergodic, continuous, measure-preserving flow on X, and
- $\epsilon > 0$.

Show that either
(a) some orbit of φ_t has measure 1, or
(b) there exist $\delta > 0$ and a compact subset K of X, such that
  - $\mu(K) > 1 - \epsilon$ and
  - $d(\varphi_t(x), x) > \delta$, for all $t \in (1 - \delta, 1 + \delta)$.

#23. Show that the maps ψ_1 and ψ_2 of Eg. 1.5.29(3) are well defined and continuous.

#24. Derive Cor. 1.5.30 from Ratner's Measure Classification Theorem.

#25. Show that if $\hat{\mu}$ is a finite-cover self-joining, then there is a $\hat{\mu}$-conull subset Ω of $X \times X$, such that $(\{x\} \times X) \cap \Omega$ and $(X \times \{x\}) \cap \Omega$ are finite, for every $x \in X$.

#26. Write a rigorous (direct) proof of Cor. 1.5.31, by choosing appropriate conull subsets of X, and so forth.
[*Hint:* You may assume (without proof) that there is a compact subset K of $X \times X$, such that $\mu(K) > 0.99$ and $K \cap \xi_s(K) = \varnothing$ for all $s \in \mathbb{R}$ with $(\xi_s)_* \hat{\mu} \neq \hat{\mu}$ (cf. 1.5.33).]

#27. Verify that $\hat{\mu}$ must be the product joining in Case 1 of the proof of Cor. 1.5.31.

#28. Suppose
- φ_t is a (measurable) flow on a measure space X,
- μ is a φ_t-invariant probability measure on X, and
- $\psi \colon X \to X$ is a Borel map that commutes with φ_t.

Show that $\psi_* \mu$ is φ_t-invariant.

#29. Suppose μ and ν are probability measures on a measure space X. Show ν has a unique decomposition $\nu = \nu_1 + \nu_2$, where $\nu_1 \perp \mu$ and $\nu_2 = f\mu$, for some $f \in L^1(\mu)$. (Recall that the notation $\mu_1 \perp \mu_2$ means the measures μ_1 and μ_2 are *singular* to each other; that is, some μ_1-conull set is μ_2-null, and vice-versa.)
[*Hint:* The map $\phi \mapsto \int \phi \, d\mu$ is a linear functional on $L^2(X, \mu +$

v), so it is represented by integration against a function $\psi \in L^2(X, \mu + v)$. Let v_1 be the restriction of v to $\psi^{-1}(0)$, and let $f = (1 - \psi)/\psi$.]

#30. Suppose
- φ_t is a (measurable) flow on a space X, and
- μ_1 and μ_2 are two different ergodic, φ_t-invariant probability measures on X.

Show that μ_1 and μ_2 are **singular** to each other.
[*Hint:* Exer. 29.]

#31. Suppose
- X is a locally compact, separable metric space,
- μ is a probability measure on X, and
- $\psi: X \to X$ is a Borel map, such that $\psi_* \mu$ and μ are singular to each other.

Show:
(a) There is a conull subset Ω of X, such that $\psi^{-1}(\Omega)$ is disjoint from Ω.
(b) For every $\epsilon > 0$, there is a compact subset K of X, such that $\mu(K) > 1 - \epsilon$ and $\psi^{-1}(K)$ is disjoint from K.

1.6. The Shearing Property for larger groups

If G is $SL(3, \mathbb{R})$, or some other group larger than $SL(2, \mathbb{R})$, then the Shearing Property is usually not true as stated in (1.5.20) or (1.5.21). This is because the centralizer of the subgroup u_t is usually larger than $\{u_t\}$.

(1.6.1) **Example.** If $y = xq$, with $q \in C_G(u_t)$, then $u^{-t}qu^t = q$ for all t, so, contrary to (1.5.21), the points x and y move together, along parallel orbits; there is no relative motion at all.

In a case where there is relative motion (that is, when $q \notin C_G(u^t)$), the fastest relative motion will usually not be along the orbits of u^t, but, rather, along some other direction in the centralizer of u^t. (We saw an example of this in the proof that self-joinings have finite fibers (see Cor. 1.5.31): under the unipotent flow $\eta_t \times \eta_t$, the points (x, a) and (y, b) move apart in the direction of the flow ξ_t, not $\eta_t \times \eta_t$.)

(1.6.2) **Proposition** (Generalized Shearing Property). *The fastest relative motion between two nearby points is along some direction in the centralizer of u^t.*
More precisely, if
- $\{u^t\}$ *is a unipotent one-parameter subgroup of G, and*

- x and y are nearby points in $\Gamma \backslash G$,

then either

1) there exists $t > 0$ and $c \in C_G(u^t)$, such that
   (a) $\|c\| = 1$, and
   (b) $xu^t \approx yu^t c$,
   or
2) there exists $c \in C_G(u^t)$, with $c \approx I$, such that $y = xc$.

Proof (*Requires some Lie theory*). Write $y = xq$, with $q \approx I$. It is easiest to work with exponential coordinates in the Lie algebra; for $g \in G$ (with g near I), let $\underline{g}$ be the (unique) small element of $\mathfrak{g}$ with $\exp \underline{g} = g$. In particular, choose

- $\underline{u} \in \mathfrak{g}$ with $\exp(t\underline{u}) = u^t$, and
- $\underline{q} \in \mathfrak{g}$ with $\exp \underline{q} = q$.

Then

$$u^{-t} \underline{q} u^t = \underline{q}(\operatorname{Ad} u^t) = \underline{q}\exp(\operatorname{ad}(t\underline{u}))$$
$$= \underline{q} + \underline{q}(\operatorname{ad}\underline{u})t + \tfrac{1}{2}\underline{q}(\operatorname{ad}\underline{u})^2 t^2 + \tfrac{1}{6}\underline{q}(\operatorname{ad}\underline{u})^3 t^3 + \cdots.$$

For large t, the largest term is the one with the highest power of t; that is, the last nonzero term $\underline{q}(\operatorname{ad}\underline{u})^k$. Then

$$[\underline{q}(\operatorname{ad}\underline{u})^k, \underline{u}] = (\underline{q}(\operatorname{ad}\underline{u})^k)(\operatorname{ad}\underline{u}) = \underline{q}(\operatorname{ad}\underline{u})^{k+1} = 0$$

(because the next term does not appear), so $\underline{q}(\operatorname{ad}\underline{u})^k$ is in the centralizer of u^t. $\qquad\square$

The above proposition shows that the direction of fastest relative motion is along the centralizer of u^t. This direction may or may not belong to $\{u^t\}$ itself. In the proof of Ratner's Theorem, it turns out that we wish to ignore motion **along** the orbits, and consider, instead, only the component of the relative motion that is **transverse** (or perpendicular) to the orbits of the flow. This direction, by definition, does not belong to $\{u^t\}$. It may or may not belong to the centralizer of $\{u^t\}$.

(1.6.3) **Example.** Assume $G = \mathrm{SL}(2, \mathbb{R})$, and suppose x and y are two points in $\Gamma \backslash G$ with $x \approx y$. Then, by continuity, $x_t \approx y_t$ for a long time. Eventually, we will be able to see a difference between x_t and y_t. The Shearing Property (1.5.20) tells us that, when this first happens, x_t will be indistinguishable from some point on the orbit of y; that is, $x_t \approx y_{t'}$ for some t'. This will continue for another long time (with t' some function of t), but we can expect that x_t will eventually diverge from the orbit of y — this is **transverse divergence**. (Note that this transverse divergence is a second-order effect; it is only apparent after we mod out the relative motion along the orbit.) Letting $y_{t'}$ be the point on the orbit of y that is closest to x_t, we write $x_t = y_{t'}g$ for some $g \in G$. Then $g - I$ represents the transverse divergence. When this transverse divergence first becomes macroscopic, we wish to understand which of the matrix entries of $g - I$ are macroscopic.

In the matrix on the RHS of Eq. (1.5.14), we have already observed that the largest entry is in the bottom left corner, the direction of $\{u^t\}$. If we ignore that entry, then the two diagonal entries are the largest entries. The diagonal corresponds to the subgroup $\{a^t\}$ (or, in geometric terms, to the geodesic flow y_t). Thus, the fastest *transverse* divergence is in the direction of $\{a^t\}$. Notice that $\{a^t\}$ normalizes $\{u^t\}$ (see Exer. 1.1#9).

(1.6.4) **Proposition.** *The fastest transverse motion is along some direction in the normalizer of u^t.*

Proof. In the calculations of the proof of Prop. 1.6.2, any term that belongs to $\mathfrak{u}$ represents motion along $\{u^t\}$. Thus, the fastest transverse motion is represented by the *last term* $\underline{q}(\operatorname{ad}\underline{u})^k$ that is *not* in $\mathfrak{u}$. Then $\underline{q}(\operatorname{ad}\underline{u})^{k+1} \in \mathfrak{u}$, or, in other words,

$$[\underline{q}(\operatorname{ad}\underline{u})^k, \underline{u}] \in \mathfrak{u}.$$

Therefore $\underline{q}(\operatorname{ad}\underline{u})^k$ normalizes $\mathfrak{u}$. □

By combining this observation with ideas from the proof that joinings have finite fibers (see Cor. 1.5.31), we see that the fastest transverse divergence is almost always in the direction of $\operatorname{Stab}_G(\mu)$, the subgroup consisting of elements of G that preserve μ. More precisely:

(1.6.5) **Corollary.** *There is a conull subset X' of X, such that, for all $x, y \in X'$, with $x \approx y$, the fastest transverse motion is along some direction in $\operatorname{Stab}_G(\mu)$.*

Proof. Because the fastest transverse motion is along the normalizer, we know that

$$yu^{t'} \approx xu^t c,$$

for some $t, t' \in \mathbb{R}$ and $c \in N_G(u^t)$.

Suppose $c \notin \operatorname{Stab}_G(\mu)$. Then, as in the proof of (1.5.31), we may assume $xu^t, yu^{t'} \in K$, where K is a large compact set, such that $K \cap Kc = \varnothing$. (Note that t' is used, instead of t, in order to eliminate relative motion *along* the $\{u^t\}$-orbit.) We have $d(K, Kc) > 0$, and this contradicts the fact that $xu^t c \approx yu^{t'}$. □

(1.6.6) **Remark.** We note an important difference between the preceding two results:

1) Proposition 1.6.4 is purely algebraic, and applies to all $x, y \in \Gamma \backslash G$ with $x \approx y$.

2) Corollary 1.6.5 depends on the measure μ — it applies only on a conull subset of $\Gamma \backslash G$.

We have considered only the case of a one-parameter subgroup $\{u^t\}$, but, for the proof of Ratner's Theorem in general, it is important to know that the analogue of Prop. 1.6.4 is also true for actions of larger unipotent subgroups U:

$$\text{the fastest transverse motion is along} \atop \text{some direction in the normalizer of } U. \qquad (1.6.7)$$

To make a more precise statement, consider two points $x, y \in X$, with $x \approx y$. When we apply larger and larger elements u of U to x, we will eventually reach a point where we can see that $xu \notin yU$. When we first reach this stage, there will be an element c of the normalizer $N_G(U)$, such that xuc appears to be in yU; that is,

$$xuc \approx yu', \text{ for some } u' \in U. \qquad (1.6.8)$$

This implies that the analogue of Cor. 1.6.5 is also true:

(1.6.9) **Corollary.** *There is a conull subset X' of X, such that, for all $x, y \in X'$, with $x \approx y$, the fastest transverse motion to the U-orbits is along some direction in* $\mathrm{Stab}_G(\mu)$.

To illustrate the importance of these results, let us prove the following special case of Ratner's Measure Classification Theorem. It is a major step forward. It shows, for example, that if μ is not supported on a single u^t-orbit, then there must be other translations in G that preserve μ.

(1.6.10) **Proposition.** *Let*

- *Γ be a lattice in a Lie group G,*
- *u^t be a unipotent one-parameter subgroup of G, and*
- *μ be an ergodic u^t-invariant probability measure on $\Gamma \backslash G$.*

If $U = \mathrm{Stab}_G(\mu)^\circ$ is unipotent, then μ is supported on a single U-orbit.

Proof. This is similar to the proof that joinings have finite fibers (see Cor. 1.5.31). We ignore some details (these may be taken to be exercises for the reader). For example, let us ignore the distinction between $\mathrm{Stab}_G(\mu)$ and its identity component $\mathrm{Stab}_G(\mu)^\circ$ (see Exer. 1).

By ergodicity, it suffices to find a U-orbit of positive measure, so let us suppose all U-orbits have measure 0. Actually, let us make the stronger assumption that all $N_G(U)$-orbits have measure 0. This will lead to a contradiction, so we can conclude that μ is supported on an orbit of $N_G(U)$. It is easy to finish from there (see Exer. 3).

By our assumption of the preceding paragraph, for almost every $x \in \Gamma\backslash G$, there exists $y \approx x$, such that

- $y \notin xN_G(U)$, and
- y is in the support of μ.

Because $y \notin x\, N_G(U)$, the U-orbit of y has nontrivial transverse divergence from the U-orbit of x (see Exer. 4), so

$$yu' \approx xuc,$$

for some $u, u' \in U$ and $c \notin U$. From Cor. 1.6.9, we know that $c \in \mathrm{Stab}_G(\mu)$. This contradicts the fact that $U = \mathrm{Stab}_G(\mu)$.   □

Exercises for §1.6.

#1. The proof we gave of Prop. 1.6.10 assumes that $\mathrm{Stab}_G(\mu)$ is unipotent. Correct the proof to use only the weaker assumption that $\mathrm{Stab}_G(\mu)^\circ$ is unipotent.

#2. Suppose

- Γ is a closed subgroup of a Lie group G,
- U is a unipotent, normal subgroup of G, and
- μ is an ergodic U-invariant probability measure on $\Gamma\backslash G$.

Show that μ is supported on a single orbit of $\mathrm{Stab}_G(\mu)$.

[*Hint:* For each $g \in N_G(U)$, such that $g \notin \mathrm{Stab}_G(\mu)$, there is a conull subset Ω of $\Gamma\backslash G$, such that $\Omega \cap g\Omega = \varnothing$ (see Exers. 1.5#30 and 1.5#31). You may assume, without proof, that this set can be chosen independent of g: there is a conull subset Ω of $\Gamma\backslash G$, such that if $g \in N_G(U)$ and $g \notin \mathrm{Stab}_G(\mu)$, then $\Omega \cap g\Omega = \varnothing$. (This will be proved in (5.8.6).)]

#3. Suppose

- Γ is a lattice in a Lie group G,
- μ is a U-invariant probability measure on $\Gamma\backslash G$, and
- μ is supported on a single $N_G(U)$-orbit.

Show that μ is supported on a single U-orbit.

[*Hint:* Reduce to the case where $N_G(U) = G$, and use Exer. 2.]

#4. In the situation of Prop. 1.6.10, show that if $x, y \in \Gamma\backslash G$, and $y \notin x\, N_G(U)$, then the U-orbit of y has nontrivial transverse divergence from the U-orbit of x.

1.7. Entropy and a proof for G = SL(2, ℝ)

The Shearing Property (and consequences such as (1.6.10)) are an important part of the proof of Ratner's Theorems, but there are two additional ingredients. We discuss the role of entropy in this section. The other ingredient, exploiting the direction of transverse divergence, is the topic of the following section.

To illustrate, let us prove Ratner's Measure Classification Theorem (1.3.7) for the case G = SL(2, ℝ):

(1.7.1) **Theorem.** *If*

- G = SL(2, ℝ),
- *Γ is any lattice in G, and*
- *η_t is the usual unipotent flow on $\Gamma \backslash G$, corresponding to the unipotent one-parameter subgroup $u^t = \begin{bmatrix} 1 & 0 \\ t & 1 \end{bmatrix}$ (see 1.1.5),*

then every ergodic η_t-invariant probability measure on $\Gamma \backslash G$ is homogeneous.

Proof. Let

- μ be any ergodic η_t-invariant probability measure on $\Gamma \backslash G$, and
- $S = \operatorname{Stab}_G(\mu)$.

We wish to show that μ is supported on a single S-orbit.

Because μ is η_t-invariant, we know that $\{u^t\} \subset S$. We may assume $\{u^t\} \neq S^\circ$. (Otherwise, it is obvious that S° is unipotent, so Prop. 1.6.10 applies.) Therefore, S° contains the diagonal one-parameter subgroup

$$a^s = \begin{bmatrix} e^s & 0 \\ 0 & e^{-s} \end{bmatrix}$$

(see Exer. 2). To complete the proof, we will show S also contains the opposite unipotent subgroup

$$v^r = \begin{bmatrix} 1 & r \\ 0 & 1 \end{bmatrix}. \tag{1.7.2}$$

Because $\{u^t\}$, $\{a^s\}$, and $\{v^r\}$, taken together, generate all of G, this implies $S = G$, so μ must be the G-invariant (Haar) measure on $\Gamma \backslash G$, which is obviously homogeneous.

Because $\{a^s\} \subset S$, we know that a^s preserves μ. Instead of continuing to exploit dynamical properties of the unipotent subgroup $\{u^t\}$, we complete the proof by working with $\{a^s\}$.

Let γ_s be the flow corresponding to a^s (see Notn. 1.1.5). The map γ_s is *not* an isometry:

- y_s multiplies infinitesimal distances in u^t-orbits by e^{2s},
- y_s multiplies infinitesimal distances in v^r-orbits by e^{-2s}, and
- y_s does act as an isometry on a^s-orbits; it multiplies infinitesimal distances along a^s-orbits by 1

(see Exer. 3). The map y_s is volume preserving because these factors cancel exactly: $e^{2s} \cdot e^{-2s} \cdot 1 = 1$.

The fact that y_s preserves the usual volume form on $\Gamma \backslash G$ led to the equation $e^{2s} \cdot e^{-2s} \cdot 1 = 1$. Let us find the analogous conclusion that results from the fact that y_s preserves the measure μ:

- Because $\{a^s\}$ normalizes $\{u^t\}$ (see Exer. 1.1#9),

$$B = \{ a^s u^t \mid s, t \in \mathbb{R} \}$$

  is a subgroup of G.
- Choose a small (2-dimensional) disk D in some B-orbit.
- For some (fixed) small $\epsilon > 0$, and each $d \in D$, let $\mathcal{B}_d = \{ dv^r \mid 0 \le r \le \epsilon \}$.
- Let $\mathcal{B} = \bigcup_{d \in D} \mathcal{B}_d$.
- Then $\mathcal{B}$ is the disjoint union of the fibers $\{\mathcal{B}_d\}_{d \in D}$, so the restriction $\mu|_{\mathcal{B}}$ can be decomposed as an integral of probability measures on the fibers:

$$\mu|_{\mathcal{B}} = \int_D \mu_d \, \nu(d),$$

  where ν_d is a probability measure on $\mathcal{B}_d$ (see 3.3.4).
- The map y_s multiplies areas in D by $e^{2s} \cdot 1 = e^{2s}$.
- Then, because μ is y_s-invariant, the contraction along the fibers $\mathcal{B}_d$ must exactly cancel this: for $X \subset \mathcal{B}_d$, we have

$$\mu_{y_s(d)}(y_s(X)) = e^{-2s}\mu_d(X).$$

The conclusion is that the fiber measures μ_d scale exactly like the Lebesgue measure on $[0, \epsilon]$. This implies, for example, that μ_d cannot be a point mass. In fact, one can use this conclusion to show that μ_d must be precisely the Lebesgue measure. (From this, it follows immediately that μ is the Haar measure on $\Gamma \backslash G$.) As will be explained below, the concept of entropy provides a convenient means to formalize the argument.   □

(1.7.3) **Notation.** As will be explained in Chap. 2, one can define the *entropy* of any measure-preserving transformation on any measure space. (Roughly speaking, it is a number that describes how quickly orbits of the transformation diverge from each other.) For any $g \in G$ and any g-invariant probability measure μ on $\Gamma \backslash G$, let $h_\mu(g)$ denote the entropy of the translation by g.

A general lemma relates entropy to the rates at which the flow expands the volume form on certain transverse foliations (see 2.5.11′). In the special case of a^s in SL(2, ℝ), it can be stated as follows.

(1.7.4) Lemma. *Suppose μ is an a^s-invariant probability measure on $\Gamma \backslash$ SL(2, ℝ).*

We have $h_\mu(a^s) \leq 2|s|$, with equality if and only if μ is $\{u^t\}$-invariant.

We also have the following general fact (see Exer. 2.3#7):

(1.7.5) Lemma. *The entropy of any invertible measure-preserving transformation is equal to the entropy of its inverse.*

Combining these two facts yields the following conclusion, which completes the proof of Thm. 1.7.1.

(1.7.6) Corollary. *Let μ be an ergodic $\{u^t\}$-invariant probability measure on $\Gamma \backslash$ SL(2, ℝ).*

If μ is $\{a^s\}$-invariant, then μ is SL(2, ℝ)-invariant.

Proof. From the equality clause of Lem. 1.7.4, we have $h_\mu(a^s) = 2|s|$, so Lem. 1.7.5 asserts that $h_\mu(a^{-s}) = 2|s|$.

On the other hand, there is an automorphism of SL(2, ℝ) that maps a^s to a^{-s}, and interchanges $\{u^t\}$ with $\{v^r\}$. Thus Lem. 1.7.4 implies:

$$h_\mu(a^{-s}) \leq 2|s|,$$

with equality if and only if μ is $\{v^r\}$-invariant.

Combining this with the conclusion of the preceding paragraph, we conclude that μ is $\{v^r\}$-invariant.

Because v^r, a^s, and u^t, taken together, generate the entire SL(2, ℝ), we conclude that μ is SL(2, ℝ)-invariant.   □

Exercises for §1.7.

#1. Let $T = \begin{bmatrix} 1 & a & c \\ 0 & 1 & b \\ 0 & 0 & 1 \end{bmatrix}$, with a, b ≠ 0. Show that if V is a vector subspace of $\mathbb{R}^3$, such that $T(V) \subset V$ and $\dim V > 1$, then $\{(0, *, 0)\} \subset V$.

#2. [*Requires some Lie theory*] Show that if H is a connected subgroup of SL(2, ℝ) that contains $\{u^t\}$ as a proper subgroup, then $\{a^s\} \subset H$.

[*Hint:* The Lie algebra of H must be invariant under $\mathrm{Ad}_G \, u^t$. For the appropriate basis of the Lie algebra $\mathfrak{sl}(2, \mathbb{R})$, the desired conclusion follows from Exer. 1.]

#3. Show:

(a) $y_s(xu^t) = y_s(x) \, u^{e^{2s}t}$,

(b) $y_s(xv^t) = y_s(x) \, v^{e^{-2s}t}$, and

(c) $y_s(xa^t) = y_s(x) \, a^t$.

1.8. Direction of divergence and a joinings proof

In §1.5, we proved only a weak form of the Joinings Theorem (1.5.30). To complete the proof of (1.5.30) and, more importantly, to illustrate another important ingredient of Ratner's proof, we provide a direct proof of the following fact:

(1.8.1) **Corollary** (Ratner). *If*

- $\hat{\mu}$ *is an ergodic self-joining of* η_t, *and*
- $\hat{\mu}$ *has finite fibers,*

then $\hat{\mu}$ *is a finite cover.*

(1.8.2) **Notation.** We fix some notation for the duration of this section. Let

- Γ be a lattice in $G = \mathrm{SL}(2, \mathbb{R})$,
- $X = \Gamma \backslash G$,
- $U = \{u^t\}$,
- $A = \{a^s\}$,
- $V = \{v^r\}$,
- $\tilde{\ }: G \to G \times G$ be the natural diagonal embedding.

At a certain point in the proof of Ratner's Measure Classification Theorem, we will know, for certain points x and $y = xg$, that the direction of fastest transverse divergence of the orbits belongs to a certain subgroup. This leads to a restriction on g. In the setting of Cor. 1.8.1, this crucial observation amounts to the following lemma.

(1.8.3) **Lemma.** *Let* $x, y \in X \times X$. *If*

- $x \approx y$,
- $y \in x(V \times V)$, *and*
- *the direction of fastest transverse divergence of the* $\tilde{U}$*-orbits of* x *and* y *belongs to* $\tilde{A}$,

then $y \in x\tilde{V}$.

Proof. We have $y = xv$ for some $v \in V \times V$. Write $x = (x_1, x_2)$, $y = (y_1, y_2)$ and $v = (v_1, v_2) = (v^{r_1}, v^{r_2})$. To determine the direction of fastest transverse divergence, we calculate

$$\widetilde{u^{-t}} v \widetilde{u^t} - (I, I) = (u^{-t} v_1 u^t - I, u^{-t} v_2 u^t - I)$$

$$\approx \left(\begin{bmatrix} r_1 t & 0 \\ -r_1 t^2 & -r_1 t \end{bmatrix}, \begin{bmatrix} r_2 t & 0 \\ -r_2 t^2 & -r_2 t \end{bmatrix} \right)$$

(cf. 1.5.14). By assumption, the largest terms of the two components must be (essentially) equal, so we conclude that $r_1 = r_2$. Therefore $v \in \hat{V}$, as desired. □

Also, as in the preceding section, the proof of Cor. 1.8.1 relies on the relation of entropy to the rates at which a flow expands the volume form on transverse foliations. For the case of interest to us here, the general lemma (2.5.11′) can be stated as follows.

(1.8.4) Lemma. *Let*

- *$\hat{\mu}$ be an $\widetilde{a^s}$-invariant probability measure on $X \times X$, and*
- *$\hat{V}$ be a connected subgroup of $V \times V$.*

Then:

1) *If $\hat{\mu}$ is $\hat{V}$-invariant, then $h_{\hat{\mu}}(\widetilde{a^s}) \geq 2|s| \dim \hat{V}$.*
2) *If there is a conull, Borel subset Ω of $X \times X$, such that $\Omega \cap x(V \times V) \subset x\hat{V}$, for every $x \in \Omega$, then $h_{\hat{\mu}}(\widetilde{a^s}) \leq 2|s| \dim \hat{V}$.*
3) *If the hypotheses of (2) are satisfied, and equality holds in its conclusion, then $\hat{\mu}$ is $\hat{V}$-invariant.*

Proof of Cor. 1.8.1. We will show that $\hat{\mu}$ is $\tilde{G}$-invariant, and is supported on a single $\tilde{G}$-orbit. (Actually, we will first replace $\tilde{G}$ by a conjugate subgroup.) Then it is easy to see that $\hat{\mu}$ is a finite cover (see Exer. 2).

It is obvious that $\hat{\mu}$ is not supported on a single $\tilde{U}$-orbit (because $\hat{\mu}$ must project to the Haar measure on each factor of $X \times X$), so, by combining (1.6.7) with (1.6.9) (and Exer. 1.6#3), we see that $\text{Stab}_{G \times G}(\hat{\mu})$ must contain a connected subgroup of $N_{G \times G}(\tilde{U})$ that is not contained in $\tilde{U}$. (Note that $N_{G \times G}(\tilde{U}) = \tilde{A} \ltimes (U \times U)$ (see Exer. 3).) Using the fact that $\hat{\mu}$ has finite fibers, we conclude that $\text{Stab}_{G \times G}(\hat{\mu})$ contains a conjugate of $\tilde{A}$ (see Exers. 4 and 5). Let us assume, without loss of generality, that $\tilde{A} \subset \text{Stab}_{G \times G}(\hat{\mu})$ (see Exer. 6); then

$$N_{G \times G}(\tilde{U}) \cap \text{Stab}_{G \times G}(\hat{\mu}) = \widetilde{AU} \qquad (1.8.5)$$

(see Exer. 7). Combining (1.6.7), (1.6.9), and (1.8.5) yields a conull subset $(X \times X)'$ of $X \times X$, such that if $x, y \in (X \times X)'$ (with $x \approx y$), then the direction of fastest transverse divergence between the $\tilde{U}$-orbits of x and y is an element of $\widetilde{AU}$. Thus, Lem. 1.8.3 implies that $(X \times X)' \cap x(V \times V) \subset x\tilde{V}$, so an entropy argument, based on Lem. 1.8.4, shows that

$$\hat{\mu} \text{ is } \tilde{V}\text{-invariant} \tag{1.8.6}$$

(see Exer. 8).

Because $\tilde{U}$, $\tilde{A}$, and $\tilde{V}$, taken together, generate $\tilde{G}$, we conclude that $\hat{\mu}$ is $\tilde{G}$-invariant. Then, because $\hat{\mu}$ has finite fibers (and is ergodic), it is easy to see that $\hat{\mu}$ is supported on a single $\tilde{G}$-orbit (see Exer. 9). $\qquad\qquad\qquad\square$

Exercises for §1.8.

#1. Obtain Cor. 1.5.9 by combining Cors. 1.5.12 and 1.8.1.

#2. In the notation of (1.8.1) and (1.8.2), show that if $\hat{\mu}$ is $\tilde{G}$-invariant, and is supported on a single $\tilde{G}$-orbit in $X \times X$, then $\hat{\mu}$ is a finite-cover joining.

 [*Hint:* The $\tilde{G}$-orbit supporting $\hat{\mu}$ can be identified with $\Gamma' \backslash G$, for some lattice Γ' in G.]

#3. In the notation of (1.8.2), show that $N_{G \times G}(\tilde{U}) = \tilde{A} \ltimes (U \times U)$.

#4. In the notation of (1.8.1), show that if $\hat{\mu}$ has finite fibers, then $\mathrm{Stab}_{G \times G}(\mu) \cap (G \times \{e\})$ is trivial.

#5. In the notation of (1.8.2), show that if H is a connected subgroup of $\tilde{A} \ltimes (U \times U)$, such that
  • $H \not\subset U \times U$, and
  • $H \cap (G \times \{e\})$ and $H \cap (\{e\} \times G)$ are trivial,

  then H contains a conjugate of $\tilde{A}$.

#6. Suppose
  • Γ is a lattice in a Lie group G,
  • μ is a measure on $\Gamma \backslash G$, and
  • $g \in G$.

  Show $\mathrm{Stab}_G(g_*\mu) = g^{-1} \mathrm{Stab}_G(\mu) g$.

#7. In the notation of (1.8.2), show that if H is a subgroup of $(A \times A) \ltimes (U \times U)$, such that
  • $\widetilde{AU} \subset H$, and
  • $H \cap (G \times \{e\})$ and $H \cap (\{e\} \times G)$ are trivial,

  then $H = \widetilde{AU}$.

#8. Establish (1.8.6).

#9. In the notation of (1.8.1) and (1.8.2), show that if $\hat{\mu}$ is $\tilde{G}$-invariant, then $\hat{\mu}$ is supported on a single $\tilde{G}$-orbit.
[*Hint:* First show that $\hat{\mu}$ is supported on a finite union of $\tilde{G}$-orbits, and then use the fact that $\hat{\mu}$ is ergodic.]

1.9. From measures to orbit closures

In this section, we sketch the main ideas used to derive Ratner's Orbit Closure Theorem (1.1.14) from her Measure Classification Theorem (1.3.7). This is a generalization of (1.3.9), and is proved along the same lines. Instead of establishing only (1.1.14), the proof yields the much stronger Equidistribution Theorem (1.3.5).

Proof of the Ratner Equidistribution Theorem. To simplify matters, let us

A) assume that $\Gamma\backslash G$ is compact, and

B) ignore the fact that not all measures are ergodic.

Remarks 1.9.1 and 1.9.3 indicate how to modify the proof to eliminate these assumptions.

Fix $x \in G$. By passing to a subgroup of G, we may assume

C) there does not exist any connected, closed, proper subgroup S of G, such that

(a) $\{u^t\}_{t\in\mathbb{R}} \subset S$,

(b) the image $[xS]$ of xS in $\Gamma\backslash G$ is closed, and has finite S-invariant volume.

We wish to show that the u^t-orbit of $[x]$ is uniformly distributed in all of $\Gamma\backslash G$, with respect to the G-invariant volume on $\Gamma\backslash G$. That is, letting

• $x_t = xu^t$ and

• $\mu_L(f) = \dfrac{1}{L}\displaystyle\int_0^L f([x_t])\,dt$,

we wish to show that the measures μ_L converge to $\text{vol}_{\Gamma\backslash G}$, as $L \to \infty$.

Assume, for simplicity, that $\Gamma\backslash G$ is compact (see A). Then the space of probability measures on $\Gamma\backslash G$ is compact (in an appropriate weak* topology), so it suffices to show that

if μ_{L_n} is any convergent sequence, then the limit μ_∞ is $\text{vol}_{\Gamma\backslash G}$.

It is easy to see that μ_∞ is u^t-invariant. Assume for simplicity, that it is also ergodic (see B). Then Ratner's Measure Classification Theorem (1.3.7) implies that there is a connected, closed subgroup S of G, and some point x' of G, such that

1) $\{u^t\}_{t\in\mathbb{R}} \subset S$,

2) the image $[x'S]$ of $x'S$ in $\Gamma\backslash G$ is closed, and has finite S-invariant volume, and

3) $\mu_\infty = \mathrm{vol}_{[x'S]}$.

It suffices to show that $[x] \in [x'S]$, for then (C) implies that $S = G$, so

$$\mu_\infty = \mathrm{vol}_{[x'S]} = \mathrm{vol}_{[x'G]} = \mathrm{vol}_{\Gamma\backslash G},$$

as desired.

To simplify the remaining details, let us assume, for the moment, that S is trivial, so μ_∞ is the point mass at the point $[x']$. (Actually, this is not possible, because $\{u^t\} \subset S$, but let us ignore this inconsistency.) This means, for any neighborhood $\mathcal{O}$ of $[x']$, no matter how small, that the orbit of $[x]$ spends more than 99% of its life in $\mathcal{O}$. By Polynomial Divergence of Orbits (cf. 1.5.18), this implies that if we enlarge $\mathcal{O}$ slightly, then the orbit is *always* in $\mathcal{O}$. Let $\widetilde{\mathcal{O}}$ be the inverse image of $\mathcal{O}$ in G. Then, for some connected component $\widetilde{\mathcal{O}}^\circ$ of $\widetilde{\mathcal{O}}$, we have $xu^t \in \widetilde{\mathcal{O}}^\circ$, for all t. But $\widetilde{\mathcal{O}}^\circ$ is a small set (it has the same diameter as its image $\mathcal{O}$ in $\Gamma\backslash G$), so this implies that xu^t is a bounded function of t. A bounded polynomial is constant, so we conclude that

$$xu^t = x \text{ for all } t \in \mathbb{R}.$$

Because $[x']$ is in the closure of the orbit of $[x]$, this implies that $[x] = [x'] \in [x'S]$, as desired.

To complete the proof, we point out that a similar argument applies even if S is not trivial. We are ignoring some technicalities, but the idea is simply that the orbit of $[x]$ must spend more than 99% of its life very close to $[x'S]$. By Polynomial Divergence of Orbits, this implies that the orbit spends all of its life fairly close to $[x'S]$. Because the distance to $[x'S]$ is a polynomial function, we conclude that it is a constant, and that this constant must be 0. So $[x] \in [x'S]$, as desired. $\square$

The following two remarks indicate how to eliminate the assumptions (A) and (B) from the proof of (1.3.5).

(1.9.1) Remark. If $\Gamma\backslash G$ is not compact, we consider its one-point compactification $X = (\Gamma\backslash G) \cup \{\infty\}$. Then

- the set $\mathrm{Prob}(X)$ of probability measures on X is compact, and

- $\mathrm{Prob}(\Gamma\backslash G) = \{ \mu \in \mathrm{Prob}(X) \mid \mu(\{\infty\}) = 0 \}$.

Thus, we need only show that the limit measure μ_∞ gives measure 0 to the point ∞. In spirit, this is a consequence of the

Polynomial Divergence of Orbits, much as in the above proof of (1.3.5), putting ∞ in the role of x'. It takes considerable ingenuity to make the idea work, but it is indeed possible. A formal statement of the result is given in the following theorem.

(1.9.2) Theorem (Dani-Margulis). *Suppose*

- Γ *is a lattice in a Lie group* G,
- u^t *is a unipotent one-parameter subgroup of* G,
- $x \in \Gamma\backslash G$,
- $\epsilon > 0$, *and*
- λ *is the Lebesgue measure on* $\mathbb{R}$.

Then there is a compact subset K of $\Gamma\backslash G$, such that

$$\limsup_{L \to \infty} \frac{\lambda\{t \in [0,L] \mid xu^t \notin K\}}{L} < \epsilon.$$

(1.9.3) Remark. Even if the limit measure μ_∞ is not ergodic, Ratner's Measure Classification Theorem tells us that each of its ergodic components is homogeneous. That is, for each ergodic component μ_z, there exist

- a point $x_z \in G$, and
- a closed, connected subgroup S_z of G,

such that

1) $\{u^t\}_{t \in \mathbb{R}} \subset S_z$,
2) the image $[x_z S_z]$ of $x_z S_z$ in $\Gamma\backslash G$ is closed, and has finite S_z-invariant volume, and
3) $\mu_z = \mathrm{vol}_{[x S_z]}$.

Arguments from algebra, based on the Borel Density Theorem, tell us that:

a) up to conjugacy, there are only countable many possibilities for the subgroups S_z (see Exer. 4.7#7), and

b) for each subgroup S_z, the point x_z must belong to a countable collection of orbits of the normalizer $N_G(S_z)$.

The ***singular set*** $\mathcal{S}(u^t)$ corresponding to u^t is the union of all of these countably many $N_G(S_z)$-orbits for all of the possible subgroups S_z. Thus:

1) $\mathcal{S}(u^t)$ is a countable union of lower-dimensional submanifolds of $\Gamma\backslash G$, and

2) if μ' is any u^t-invariant probability measure on $\Gamma\backslash G$, such that $\mu'(\mathcal{S}(u^t)) = 0$, then μ' is the Lebesgue measure.

So we simply wish to show that $\mu_\infty(\mathscr{S}(u^t)) = 0$.

This conclusion follows from the polynomial speed of unipotent flows. Indeed, for every $\epsilon > 0$, because $x \notin \mathscr{S}(u^t)$, one can show there is an open neighborhood $\mathscr{O}$ of $\mathscr{S}(u^t)$, such that

$$\frac{\lambda\{t \in [0,L] \mid xu^t \in \mathscr{O}\}}{L} < \epsilon \quad \text{for every } L > 0, \qquad (1.9.4)$$

where λ is the Lebesgue measure on $\mathbb{R}$.

For many applications, it is useful to have the following stronger ("uniform") version of the Equidistribution Theorem (1.3.4):

(1.9.5) Theorem. *Suppose*

- *Γ is a lattice in a connected Lie group G,*
- *μ is the G-invariant probability measure on $\Gamma \backslash G$,*
- *$\{u_n^t\}$ is a sequence of one-parameter unipotent subgroups of G, converging to a one-parameter subgroup u^t (that is, $u_n^t \to u^t$ for all t),*
- *$\{x_n\}$ is a convergent sequence of points in $\Gamma \backslash G$, such that $\lim_{n\to\infty} x_n \notin \mathscr{S}(u^t)$,*
- *$\{L_n\}$ is a sequence of real numbers tending to ∞, and*
- *f is any bounded, continuous function on $\Gamma \backslash G$.*

Then

$$\lim_{n\to\infty} \frac{1}{L_n} \int_0^{L_n} f(x_n u_n^t)\, dt = \int_{\Gamma \backslash G} f\, d\mu.$$

Exercises for §1.9.

#1. Reversing the logical order, prove that Thm. 1.9.2 is a corollary of the Equidistribution Theorem (1.3.4).

#2. Suppose S is a subgroup of G, and H is a subgroup of S. Show, for all $g \in N_G(S)$ and all $h \in H$, that $Sgh = Sg$.

Brief history of Ratner's Theorems

In the 1930's, G. Hedlund [21, 22, 23] proved that if $G = \mathrm{SL}(2, \mathbb{R})$ and $\Gamma \backslash G$ is compact, then unipotent flows on $\Gamma \backslash G$ are ergodic and minimal.

It was not until 1970 that H. Furstenberg [19] proved these flows are uniquely ergodic, thus establishing the Measure Classification Theorem for this case. At about the same time, W. Parry [37, 38] proved an Orbit Closure Theorem, Measure Classification Theorem, and Equidistribution Theorem for the case where G is nilpotent, and G. A. Margulis [28] used the polynomial speed of unipotent flows to prove the important fact that unipotent orbits cannot go off to infinity.

Inspired by these and other early results, M. S. Raghunathan conjectured a version of the Orbit Closure Theorem, and showed that it would imply the Oppenheim Conjecture. Apparently, he did not publish this conjecture, but it appeared in a paper of S. G. Dani [7] in 1981. In this paper, Dani conjectured a version of the Measure Classification Theorem.

Dani [6] also generalized Furstenberg's Theorem to the case where $\Gamma \backslash SL(2, \mathbb{R})$ is not compact. Publications of R. Bowen [4], S. G. Dani [6, 7], R. Ellis and W. Perrizo [17], and W. Veech [60] proved further generalizations for the case where the unipotent subgroup U is horospherical (see 2.5.6 for the definition). (Results for horosphericals also follow from a method in the thesis of G. A. Margulis [27, Lem. 5.2] (cf. Exer. 5.7#5).)

M. Ratner began her work on the subject at about this time, proving her Rigidity Theorem (1.5.3), Quotients Theorem (1.5.9), Joinings Theorem (1.5.30), and other fundamental results in the early 1980's [41, 42, 43]. (See [44] for an overview of this early work.) Using Ratner's methods, D. Witte [61, 62] generalized her rigidity theorem to all G.

S. G. Dani and J. Smillie [16] proved the Equidistribution Theorem when $G = SL(2, \mathbb{R})$. S. G. Dani [8] showed that unipotent orbits spend only a negligible fraction of their life near infinity. A. Starkov [57] proved an orbit closure theorem for the case where G is solvable.

Using unipotent flows, G. A. Margulis' [29] proved the Oppenheim Conjecture (1.2.2) on values of quadratic forms. He and S. G. Dani [12, 13, 14] then proved a number of results, including the first example of an orbit closure theorem for actions of non-horospherical unipotent subgroups of a semisimple Lie group — namely, for "generic" one-parameter unipotent subgroups of $SL(3, \mathbb{R})$. (G. A Margulis [32, §3.8, top of p. 319] has pointed out that the methods could yield a proof of the general case of the Orbit Closure Theorem.)

Then M. Ratner [45, 46, 47, 48] proved her amazing theorems (largely independently of the work of others), by expanding the ideas from her earlier study of horocycle flows. (In the meantime, N. Shah [55] showed that the Measure Classification Theorem implied an Equidistribution Theorem for many cases when $G = SL(3, \mathbb{R})$.)

Ratner's Theorems were soon generalized to p-adic groups, by M. Ratner [51] and, independently, by G. A. Margulis and

G. Tomanov [33, 34]. N. Shah [56] generalized the results to subgroups generated by unipotent elements (1.1.19). (For **connected** subgroups generated by unipotent elements, this was proved in Ratner's original papers.)

Notes

§1.1. See [2] for an excellent introduction to the general area of flows on homogeneous spaces. Surveys at an advanced level are given in [9, 11, 24, 31, 58]. Discussions of Ratner's Theorems can be found in [2, 9, 20, 50, 52, 58].

Raghunathan's book [40] is the standard reference for basic properties of lattice subgroups.

The dynamical behavior of the geodesic flow can be studied by associating a continued fraction to each point of $\Gamma \backslash G$. (See [1] for an elementary explanation.) In this representation, the fact that some orbit closures are fractal sets (1.1.9) is an easy observation.

See [40, Thms. 1.12 and 1.13, pp. 22-23] for solutions of Exers. 1.1#23 and 1.1#24.

§1.2. Margulis' Theorem on values of quadratic forms (1.2.2) was proved in [29], by using unipotent flows. For a discussion and history of this theorem, and its previous life as the Oppenheim Conjecture, see [32]. An elementary proof is given in [14], [2, Chap. 6], and [10].

§1.3. M. Ratner proved her Measure Classification Theorem (1.3.7) in [45, 46, 47]. She then derived her Equidistribution Theorem (1.3.4) and her Orbit Closure Theorem (1.1.14) in [48]. A derivation also appears in [15], and an outline can be found in [33, §11].

In her original proof of the Measure Classification Theorem, Ratner only assumed that Γ is **discrete**, not that it is a lattice. D. Witte [63, §3] observed that discreteness is also not necessary (Rem. 1.3.10(1)).

See [39, §12] for a discussion of Choquet's Theorem, including a solution of Exer. 1.3#6.

§1.4. The quantitative version (1.4.1) of Margulis' Theorem on values of quadratic forms is due to A. Eskin, G. A. Margulis, and S. Mozes [18]. See [32] for more discussion of the proof, and the partial results that preceded it.

See [26] for a discussion of Quantum Unique Ergodicity and related results. Conjecture 1.4.6 (in a more general form) is due to Z. Rudnick and P. Sarnack. Theorem 1.4.8 was proved by E. Lindenstrauss [26]. The crucial fact that $h_{\hat{\mu}}(a_t) \neq 0$ was proved by J. Bourgain and E. Lindenstrauss [3].

The results of §1.4C are due to H. Oh [35].

§1.5. The Ratner Rigidity Theorem (1.5.3) was proved in [41]. Remark 1.5.4(1) was proved in [61, 62].

Flows by diagonal subgroups were proved to be Bernoulli (see 1.5.4(2)) by S. G. Dani [5], using methods of D. Ornstein and B. Weiss [36].

The Ratner Quotients Theorem (1.5.9) was proved in [42], together with Cor. 1.5.12.

The crucial property (1.5.18) of polynomial divergence was introduced by M. Ratner [41, §2] for unipotent flows on homogeneous spaces of $SL(2, \mathbb{R})$. Similar ideas had previously been used by Margulis in [28] for more general unipotent flows.

The Shearing Property (1.5.20 and 1.5.21) was introduced by M. Ratner [42, Lem. 2.1] in the proof of her Quotients Theorem (1.5.9), and was also a crucial ingredient in the proof [43] of her Joinings Theorem (1.5.30). She [43, Defn. 1] named a certain precise version of this the "H-property," in honor of the horocycle flow.

An introduction to Nonstandard Analysis (the rigorous theory of infinitesimals) can be found in [53] or [59].

Lusin's Theorem (Exer. 1.5#21) and the decomposition of a measure into a singular part and an absolutely continuous part (see Exer. 1.5#29) appear in many graduate analysis texts, such as [54, Thms. 2.23 and 6.9].

§1.6. The generalization (1.6.2) of the Shearing Property to other Lie groups appears in [61, §6], and was called the "Ratner property." The important extension (1.6.7) to transverse divergence of actions of higher-dimensional unipotent subgroups is implicit in the "R-property" introduced by M. Ratner [45, Thm. 3.1]. In fact, the R-property combines (1.6.7) with polynomial divergence. It played an essential role in Ratner's proof of the Measure Classification Theorem.

The arguments used in the proofs of (1.6.5) and (1.6.10) appear in [49, Thm. 4.1].

§1.7. Theorem 1.7.1 was proved by S. G. Dani [6], using methods of H. Furstenberg [19]. Elementary proofs based on Ratner's ideas (without using entropy) can be found in [49], [2, §4.3], [20], and [58, §16].

The entropy estimates (1.7.4) and (1.8.4) are special cases of a result of G. A. Margulis and G. Tomanov [33, Thm. 9.7]. (Margulis and Tomanov were influenced by a theorem of F. Ledrappier and L.-S. Young [25].) We discuss the Margulis-Tomanov result in §2.5, and give a sketch of the proof in §2.6.

§1.8. The subgroup $\tilde{V}$ will be called $\tilde{S}_-$ in §5.4. The proof of Lem. 1.8.3 essentially amounts to a verification of Eg. 5.4.3(5).

The key point (1.8.5) in the proof of Cor. 1.8.1 is a special case of Prop. 5.6.1.

§1.9. G. A. Margulis [28] proved a weak version of Thm. 1.9.2 in 1971. Namely, for each $x \in \Gamma \backslash G$, he showed there is a compact subset K of $\Gamma \backslash G$, such that

$$\{\, t \in [0, \infty) \mid [xu^t] \in K \,\} \tag{1.10.6}$$

is unbounded. The argument is elementary, but ingenious. A very nice version of the proof appears in [14, Appendix] (and [2, §V.5]).

Fifteen years later, S. G. Dani [8] refined Margulis' proof, and obtained (1.9.2), by showing that the set (1.10.6) not only is unbounded, but has density $> 1 - \epsilon$. The special case of Thm. 1.9.2 in which $G = \mathrm{SL}(2, \mathbb{R})$ and $\Gamma = \mathrm{SL}(2, \mathbb{Z})$ can be proved easily (see [49, Thm. 3.1] or [58, Thm. 12.2, p. 96]).

The uniform version (1.9.5) of the Equidistribution Theorem was proved by S. G. Dani and G. A. Margulis [15]. The crucial inequality (1.9.4) is obtained from the Dani-Margulis *linearization method* introduced in [15, §3].

References

[1] P. Arnoux: Le codage du flot géodésique sur la surface modulaire. *Enseign. Math.* 40 (1994), no. 1-2, 29–48. MR 95c:58136

[2] B. Bekka and M. Mayer: *Ergodic Theory and Topological Dynamics of Group Actions on Homogeneous Spaces.* London Math. Soc. Lec. Notes #269. Cambridge U. Press, Cambridge, 2000. ISBN 0-521-66030-0, MR 2002c:37002

[3] J. Bourgain and E. Lindenstrauss: Entropy of quantum limits. *Comm. Math. Phys.* 233 (2003), no. 1, 153–171. MR 2004c:11076

[4] R. Bowen: Weak mixing and unique ergodicity on homogeneous spaces. *Israel J. Math* 23 (1976), 267–273. MR 53 #11016

[5] S. G. Dani: Bernoullian translations and minimal horospheres on homogeneous spaces. *J. Indian Math. Soc.* (N.S.) 40 (1976), no. 1-4, 245–284. MR 57 #585

[6] S. G. Dani: Invariant measures of horospherical flows on noncompact homogeneous spaces. *Invent. Math.* 47 (1978), no. 2, 101–138. MR 58 #28260

[7] S. G. Dani: Invariant measures and minimal sets of horospherical flows. *Invent. Math.* 64 (1981), 357–385. MR 83c:22009

[8] S. G. Dani: On orbits of unipotent flows on homogeneous spaces II. *Ergodic Th. Dyn. Sys.* 6 (1986), no. 2, 167–182; correction 6 (1986), no. 2, 321. MR 88e:58052, MR 88e:58053

[9] S. G. Dani: Flows on homogeneous spaces: a review, in: M. Pollicott and K. Schmidt, eds., *Ergodic Theory of $\mathbb{Z}^d$ Actions* (Warwick, 1993–1994), 63–112, London Math. Soc. Lecture Note Ser., 228, Cambridge Univ. Press, Cambridge, 1996. ISBN 0-521-57688-1, MR 98b:22023

[10] S. G. Dani: On the Oppenheim conjecture on values of quadratic forms, in: É. Ghys et al., eds., *Essays on Geometry and Related Topics,* Vol. 1, pp. 257–270. Monogr. Enseign. Math. #38. Enseignement Math., Geneva, 2001. ISBN 2-940264-05-8, MR 2003m:11100

[11] S. G. Dani: Dynamical systems on homogeneous spaces, in: Ya. G. Sinai, ed., *Dynamical Systems, Ergodic Theory and*

Applications, 2nd ed. pp. 264–359. Encyclopaedia of Mathematical Sciences #100. Springer-Verlag, Berlin, 2000. ISBN 3-540-66316-9, MR 2001k:37004

[12] S. G. Dani and G. A. Margulis: Values of quadratic forms at primitive integral points. *Invent. Math.* 98 (1989), 405–424. MR 90k:22013b

[13] S. G. Dani and G. A. Margulis: Orbit closures of generic unipotent flows on homogeneous spaces of SL(3, ℝ). *Math. Ann.* 286 (1990), 101–128. MR 91k:22026

[14] S. G. Dani and G. A. Margulis: Values of quadratic forms at integral points: an elementary approach. *Enseign. Math.* 36 (1990), 143–174. MR 91k:11053

[15] S. G. Dani and G. A. Margulis: Limit distributions of orbits of unipotent flows and values of quadratic forms. *Adv. Soviet Math.* 16, Part 1, 91–137. *I. M. Gel'fand Seminar*, Amer. Math. Soc., Providence, RI, 1993. MR 95b:22024

[16] S. G. Dani and J. Smillie: Uniform distribution of horocycle orbits for Fuchsian groups. *Duke Math. J.* 51 (1984), no. 1, 185–194. MR 85f:58093

[17] R. Ellis and W. Perrizo: Unique ergodicity of flows on homogeneous spaces. *Israel J. Math.* 29 (1978), 276–284. MR 57 #12774

[18] A. Eskin, G. A. Margulis, and S. Mozes: Upper bounds and asymptotics in a quantitative version of the Oppenheim conjecture. *Ann. of Math.* 147 (1998), no. 1, 93–141. MR 99a:11043

[19] H. Furstenberg: The unique ergodicity of the horocycle flow, in: A. Beck, ed., *Recent Advances in Topological Dynamics.* Springer Lecture Notes 318 (1973), 95–115. ISBN 0-387-061878, MR 52 #14149

[20] É. Ghys: Dynamique des flots unipotents sur les espaces homogènes. Séminaire Bourbaki, vol. 1991/92. *Astérisque* 206 (1992), no. 747, 3, 93–136. MR 94e:58101

[21] G. A. Hedlund: Fuchsian groups and transitive horocycles. *Duke J. Math.* 2 (1936), 530–542. JFM 62.0392.03, Zbl 0015.10201

[22] G. A. Hedlund: The dynamics of geodesic flows. *Bull. Amer. Math. Soc.* 45 (1939), 241–260. JFM 65.0793.02

[23] G. A. Hedlund: Fuchsian groups and mixtures. *Ann. of Math.* 40 (1939), 370–383. MR 1503464, JFM 65.0793.01

[24] D. Kleinbock, N. Shah, and A. Starkov: Dynamics of subgroup actions on homogeneous spaces of Lie groups and applications to number theory, in: B. Hasselblatt and A. Katok, eds., *Handbook of Dynamical Systems, Vol. 1A,* pp. 813–930. North-Holland, Amsterdam, 2002. ISBN 0-444-82669-6, MR 2004b:22021

[25] F. Ledrappier and L.-S. Young: The metric entropy of diffeomorphisms. I. Characterization of measures satisfying Pesin's entropy formula. *Ann. of Math.* 122 (1985), no. 3, 509–539. MR 87i:58101a

[26] E. Lindenstrauss: Invariant measures and arithmetic quantum unique ergodicity. *Ann. of Math.* (to appear).

[27] G. A. Margulis: *On Some Aspects of the Theory of Anosov Systems.* Springer, Berlin, 2004. ISBN 3-540-40121-0, MR 2035655

[28] G. A. Margulis: On the action of unipotent groups in the space of lattices, in: I. M. Gel'fand, ed., *Lie Groups and Their Representations (Proc. Summer School, Bolyai, János Math. Soc., Budapest, 1971),* pp. 365–370. Halsted, New York, 1975. (Also *Math. USSR-Sb.* 15 (1972), 549–554.) MR 57 #9907, MR 45 #445

[29] G. A. Margulis: Formes quadratiques indéfinies et flots unipotents sur les espaces homogènes. *C. R. Acad. Sci. Paris Sér. I Math.* 304 (1987), no. 10, 249–253. MR 88f:11027

[30] G. A. Margulis: Discrete subgroups and ergodic theory, in: K. E. Aubert, E. Bombieri and D. Goldfeld, eds., *Number Theory, Trace Formulas and Discrete Groups (Oslo, 1987),* pp. 377–398. Academic Press, Boston, MA, 1989. ISBN: 0-12-067570-6, MR 90k:22013a

[31] G. A. Margulis: Dynamical and ergodic properties of subgroup actions on homogeneous spaces with applications to number theory, in: I. Satake, ed., *Proc. Internat. Cong. Math.,* Kyoto, 1990 (Springer, 1991), pp. 193-215. ISBN 4-431-70047-1, MR 93g:22011

[32] G. A. Margulis: Oppenheim conjecture, in: M. Atiyah and D. Iagolnitzer, eds., *Fields Medallists' Lectures,* pp. 272–327. World Sci. Ser. 20th Century Math., 5, World Sci. Publishing, River Edge, NJ, 1997. ISBN: 981-02-3117-2, MR 99e:11046

[33] G. A. Margulis and G. M. Tomanov: Invariant measures for actions of unipotent groups over local fields on homogeneous spaces. *Invent. Math.* 116 (1994), 347–392. (Announced in *C. R. Acad. Sci. Paris Sér. I Math.* 315 (1992), no. 12, 1221–1226.) MR 94f:22016, MR 95k:22013

[34] G. A. Margulis and G. M. Tomanov: Measure rigidity for almost linear groups and its applications. *J. Anal. Math.* 69 (1996), 25–54. MR 98i:22016

[35] H. Oh: Discrete subgroups generated by lattices in opposite horospherical subgroups. *J. Algebra* 203 (1998), no. 2, 621–676. MR 99b:22021

[36] D. S. Ornstein and B. Weiss: Geodesic flows are Bernoullian. *Israel J. Math.* 14 (1973), 184–198. MR 48 #4272

[37] W. Parry: Ergodic properties of affine transformations and flows on nilmanifolds. *Amer. J. Math.* 91 (1969), 757–771. MR 41 #5595

[38] W. Parry: Metric classification of ergodic nilflows and unipotent affines. *Amer. J. Math.* 93 (1971), 819–828. MR 44 #1792

[39] R. R. Phelps: *Lectures on Choquet's Theorem, 2nd ed.* Lecture Notes in Mathematics #1757. Springer-Verlag, Berlin, 2001. ISBN 3-540-41834-2, MR 2002k:46001

[40] M. S. Raghunathan: *Discrete subgroups of Lie groups.* Springer-Verlag, New York, 1972. ISBN 0-387-05749-8, MR 58 #22394a

[41] M. Ratner: Rigidity of horocycle flows. *Ann. of Math.* 115 (1982), 597–614. MR 84e:58062

[42] M. Ratner: Factors of horocycle flows. *Ergodic Theory Dyn. Syst.* 2 (1982), 465–489. MR 86a:58076

[43] M. Ratner: Horocycle flows, joinings and rigidity of products. *Ann. of Math.* 118 (1983), 277–313. MR 85k:58063

[44] M. Ratner: Ergodic theory in hyperbolic space. *Contemporary Math.* 26 (1984), 309–334. MR 85h:58140

[45] M. Ratner: Strict measure rigidity for unipotent subgroups of solvable groups. *Invent. Math.* 101 (1990), 449–482. MR 92h:22015

[46] M. Ratner: On measure rigidity of unipotent subgroups of semisimple groups. *Acta Math.* 165 (1990), 229–309. MR 91m:57031

[47] M. Ratner: On Raghunathan's measure conjecture. *Ann. of Math.* 134 (1991), 545–607. MR 93a:22009

[48] M. Ratner: Raghunathan's topological conjecture and distributions of unipotent flows. *Duke Math. J.* 63 (1991), no. 1, 235–280. MR 93f:22012

[49] M. Ratner: Raghunathan's conjectures for SL(2, ℝ). *Israel J. Math.* 80 (1992), 1–31. MR 94k:22024

[50] M. Ratner: Interactions between ergodic theory, Lie groups, and number theory, in: S. D. Chatterji, ed., *Proc. Internat. Cong. Math.*, Vol. 1 (Zürich, 1994), pp. 157–182, Birkhäuser, Basel, 1995. ISBN 3-7643-5153-5, MR 98k:22046

[51] M. Ratner: Raghunathan's conjectures for Cartesian products of real and p-adic Lie groups. *Duke Math. J.* 77 (1995), no. 2, 275–382. MR 96d:22015

[52] M. Ratner: On the p-adic and S-arithmetic generalizations of Raghunathan's conjectures, in: S. G. Dani, ed., *Lie Groups and Ergodic Theory (Mumbai, 1996)*, Narosa, New Delhi, 1998, pp. 167–202. ISBN 81-7319-235-9, MR 2001f:22058

[53] A. Robinson: *Non-Standard Analysis, 2nd ed.* Princeton University Press, Princeton, NJ, 1996. ISBN 0-691-04490-2, MR 96j:03090

[54] W. Rudin: *Real and Complex Analysis, 2nd ed.* McGraw-Hill, New York, 1974. MR 49 #8783, ISBN 0-07-054233-3

[55] N. Shah: Uniformly distributed orbits of certain flows on homogeneous spaces. *Math. Ann.* 289 (1991), 315–334. MR 93d:22010

[56] N. A. Shah: Invariant measures and orbit closures on homogeneous spaces for actions of subgroups generated by unipotent elements. in: S. G. Dani, ed., *Lie Groups and Ergodic Theory (Mumbai, 1996)*, Narosa, New Delhi, 1998, pp. 229–271. ISBN 81-7319-235-9, MR 2001a:22012

[57] A. Starkov: Solvable homogeneous flows. *Math. USSR-Sb.* 62 (1989), 243–260. MR 89b:22009

[58] A. Starkov: *Dynamical Systems on Homogeneous Spaces.* Translations of Mathematical Monographs #190. American Mathematical Society, Providence, 2000. ISBN 0-8218-1389-7, MR 2001m:37013b

[59] K. D. Stroyan and W. A. J. Luxemburg: *Introduction to the Theory of Infinitesimals.* Academic Press, New York, 1976. ISBN 0-12-674150-6, MR 58 #10429

[60] W. A. Veech: Unique ergodicity of horospherical flows. *Amer. J. Math.* 99 (1977), 827–859. MR 56 #5788

[61] D. Witte: Rigidity of some translations on homogeneous spaces. *Invent. Math.* 81 (1985), no. 1, 1–27. MR 87d:22018

[62] D. Witte: Zero-entropy affine maps. *Amer. J. Math.* 109 (1987), no. 5, 927–961; correction 118 (1996), no. 5, 1137–1140. MR 88i:28038, MR 97j:28043

[63] D. Witte: Measurable quotients of unipotent translations on homogeneous spaces. *Trans. Amer. Math. Soc.* 345 (1994), no. 2, 577–594. Correction and extension 349 (1997), no. 11, 4685–4688. MR 95a:22005, MR 98c:22002

CHAPTER 2

Introduction to Entropy

The entropy of a dynamical system can be intuitively described as a number that expresses the amount of "unpredictability" in the system. Before beginning the formal development, let us illustrate the idea by contrasting two examples.

2.1. Two dynamical systems

(2.1.1) **Definition.** In classical ergodic theory, a *dynamical system* (with discrete time) is an action of $\mathbb{Z}$ on a measure space, or, in other words, a measurable transformation $T: \Omega \to \Omega$ on a measure space Ω. (The intuition is that the points of Ω are in motion. A particle that is at a certain point $\omega \in \Omega$ will move to a point $T(\omega)$ after a unit of time.) We assume:

    1) T has a (measurable) inverse $T^{-1}: \Omega \to \Omega$, and

    2) there is a T-invariant probability measure μ on Ω.

The assumption that μ is T-*invariant* means $\mu(T(A)) = \mu(A)$, for every measurable subset A of Ω. (This generalizes the notion of an *incompressible fluid* in fluid dynamics.)

(2.1.2) **Example** (Irrational rotation of the circle). Let $\mathbb{T} = \mathbb{R}/\mathbb{Z}$ be the circle group; for any $\beta \in \mathbb{R}$, we have a measurable transformation $T_\beta: \mathbb{T} \to \mathbb{T}$ defined by

$$T_\beta(t) = t + \beta.$$

(The usual arc-length Lebesgue measure is T_β-invariant.) In physical terms, we have a circular hoop of circumference 1 that is rotating at a speed of β (see Fig. 2.1A). Note that we are taking the *circumference,* not the radius, to be 1.

If β is irrational, it is well known that every orbit of this dynamical system is uniformly distributed in $\mathbb{T}$ (see Exer. 2) (so the dynamical system is *uniquely ergodic*).

(2.1.3) **Example** (Bernoulli shift). Our other basic example comes from the study of coin tossing. Assuming we have a fair coin, this is modeled by a two event universe $C = \{H, T\}$, in which each event has probability $1/2$. The probability space for tossing two

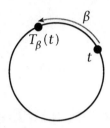

FIGURE 2.1A. T_β moves each point on the the circle a distance of β. In other words, T_β rotates the
circle 360β degrees.

coins (independently) is the product space $C \times C$, with the product measure (each of the four possible events has probability
$(1/2)^2 = 1/4$). For n tosses, we take the product measure on
C^n (each of the 2^n possible events has probability $(1/2)^n$). Now
consider tossing a coin once each day for all eternity (this is a
doubly infinite sequence of coin tosses). The probability space
is an infinite cartesian product

$$C^\infty = \{ f : \mathbb{Z} \to C \}$$

with the **product measure**: for any two disjoint finite subsets $\mathscr{H}$
and $\mathscr{T}$ of $\mathbb{Z}$, the probability that

- $f(n) = \mathsf{H}$ for all $n \in \mathscr{H}$, and
- $f(n) = \mathsf{T}$ for all $n \in \mathscr{T}$

is exactly $(1/2)^{|\mathscr{H}|+|\mathscr{T}|}$.

A particular coin-tossing history is represented by a single
element $f \in C^\infty$. Specifically, $f(0)$ is the result of today's coin
toss, $f(n)$ is the result of the toss n days from now (assuming
$n > 0$), and $f(-n)$ is the result of the toss n days ago. Tomorrow, the history will be represented by an element $g \in C^\infty$ that is
closely related to f, namely, $f(n+1) = g(n)$ for every n. (Saying
today that "I will toss a head $n + 1$ days from now" is equivalent
to saying *tomorrow* that "I will toss a head n days from now.")
This suggests a dynamical system (called a **Bernoulli shift**) that
consists of translating each sequence of H's and T's one notch
to the left:

$$T_{\text{Bern}} : C^\infty \to C^\infty \text{ is defined by } (T_{\text{Bern}}f)(n) = f(n + 1).$$

It is well known that almost every coin-tossing history consists
(in the limit) of half heads and half tails. More generally, it can
be shown that almost every orbit in this dynamical system is
uniformly distributed, so it is "**ergodic**."

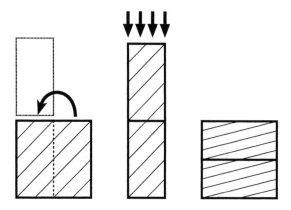

FIGURE 2.1B. Baker's Transformation: the right half of the loaf is placed on top of the left, and then the pile is pressed down to its original height.

It is also helpful to see the Bernoulli shift from a more concrete point of view:

(2.1.4) **Example** (Baker's Transformation). A baker starts with a lump of dough in the shape of the unit square $[0,1]^2$. She cuts the dough in half along the line $x = 1/2$ (see Fig. 2.1B), and places the right half $[1/2,1] \times [0,1]$ above the left half (to become $[0,1/2] \times [1,2]$). The dough is now 2 units tall (and $1/2$ unit wide). She then pushes down on the dough, reducing its height, so it widens to retain the same area, until the dough is again square. (The pushing applies the linear map $(x,y) \mapsto (2x, y/2)$ to the dough.) More formally,

$$T_{\text{Bake}}(x,y) = \begin{cases} (2x, y/2) & \text{if } x \leq 1/2 \\ (2x - 1, (y + 1)/2) & \text{if } x \geq 1/2. \end{cases} \qquad (2.1.5)$$

(This is not well defined on the set $\{x = 1/2\}$ of measure zero.)

Any point (x,y) in $[0,1]^2$ can be represented, in binary, by two strings of 0's and 1's:

$$(x,y) = (0.x_0 x_1 x_2 \ldots, 0.y_1 y_2 y_3 \ldots),$$

and we have

$$T_{\text{Bake}}(0.x_0 x_1 x_2 \ldots, 0.y_1 y_2 y_3 \ldots) = (0.x_1 x_2 \ldots, 0.x_0 y_1 y_2 y_3 \ldots),$$

so we see that

$$\begin{array}{cc} T_{\text{Bern}} \text{ is isomorphic to } T_{\text{Bake}} \\ \text{(modulo a set of measure zero),} \end{array} \qquad (2.1.6)$$

by identifying $f \in C^\infty$ with

$$(0.\widehat{f(0)}\, \widehat{f(1)}\, \widehat{f(2)}\dots, 0.\widehat{f(-1)}\, \widehat{f(-2)}\, \widehat{f(-3)}\dots),$$

where $\hat{\mathsf{H}} = 0$ and $\hat{\mathsf{T}} = 1$ (or vice-versa).

Exercises for §2.1.

#1. Show β is rational if and only if there is some positive integer k, such that $(T_\beta)^k(x) = x$ for all $x \in \mathbb{T}$.

#2. Show that if β is irrational, then every orbit of T_β is uniformly distributed on the circle; that is, if I is any arc of the circle, and x is any point in $\mathbb{T}$, show that

$$\lim_{N\to\infty} \frac{\#\{k \in \{1,2,\dots,N\} \mid T_\beta^k(x) \in I\}}{N} = \mathrm{length}(I).$$

[*Hint:* Exer. 1.3#1.]

2.2. Unpredictability

There is a fundamental difference between our two examples: the Bernoulli shift (or Baker's Transformation) is much more "random" or "unpredictable" than an irrational rotation. (Formally, this will be reflected in the fact that the entropy of a Bernoulli shift is nonzero, whereas that of an irrational rotation is zero.) Both of these dynamical systems are deterministic, so, from a certain point of view, neither is unpredictable. But the issue here is to predict behavior of the dynamical system from imperfect knowledge, and these two examples look fundamentally different from this point of view.

(2.2.1) **Example.** Suppose we have know the location of some point x of $\mathbb{T}$ to within a distance of less than 0.1; that is, we have a point $x_0 \in \mathbb{T}$, and we know that $d(x,x_0) < 0.1$. Then, for every n, we also know the location of $T_\beta^n(x)$ to within a distance of 0.1. Namely, $d(T_\beta^n(x), T_\beta^n(x_0)) < 0.1$, because T_β is an isometry of the circle. Thus, we can predict the location of $T_\beta^n(x)$ fairly accurately.

The Baker's Transformation is *not* predictable in this sense:

(2.2.2) **Example.** Suppose there is an impurity in the baker's bread dough, and we know its location to within a distance of less than 0.1. After the dough has been kneaded once, our uncertainty in the horizontal direction has doubled (see Fig. 2.2A). Kneading again doubles the horizontal uncertainty, and perhaps adds a second possible vertical location (if the cut $\{x = 1/2\}$

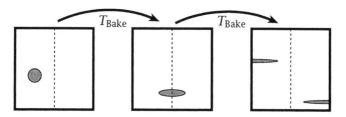

FIGURE 2.2A. Kneading the dough stretches a circular disk to a narrow, horizontal ellipse.

goes through the region where the impurity might be). As the baker continues to knead, our hold on the impurity continues to deteriorate (very quickly — at an exponential rate!). After, say, 20 kneadings, the impurity could be almost anywhere in the loaf — every point in the dough is at a distance less than 0.001 from a point where the impurity could be (see Exer. 1). In particular, we now have no idea whether the impurity is in the left half of the loaf or the right half.

The upshot is that a small change in an initial position can quickly lead to a large change in the subsequent location of a particle. Thus, errors in measurement make it impossible to predict the future course of the system with any reasonable precision. (Many scientists believe that weather forecasting suffers from this difficulty — it is said that a butterfly flapping its wings in Africa may affect the next month's rainfall in Chicago.)

To understand entropy, it is important to look at unpredictability from a different point of view, which can easily be illustrated by the Bernoulli shift.

(2.2.3) **Example.** Suppose we have tossed a fair coin 1000 times. Knowing the results of these tosses does not help us predict the next toss: there is still a 50% chance of heads and a 50% chance of tails.

More formally, define a function $\chi \colon C^\infty \to \{H, T\}$ by $\chi(f) = f(0)$. Then the values

$$\chi(T^{-1000}(f)), \chi(T^{-999}(f)), \chi(T^{-998}(f)), \ldots, \chi(T^{-1}(f))$$

give no information about the value of $\chi(f)$. Thus, T_{Bern} is quite unpredictable, even if we have a lot of past history to go on. As we will see, this means that the entropy of T_{Bern} is not zero.

(2.2.4) **Example.** In contrast, consider an irrational rotation T_β. For concreteness, let us take $\beta = \sqrt{3}/100 = .01732\ldots$, and, for convenience, let us identify $\mathbb{T}$ with the half-open interval $[0, 1)$

(in the natural way). Let $\chi: [0,1) \to \{0,1\}$ be the characteristic function of $[1/2, 1)$. Then, because $\beta \approx 1/60$, the sequence

$$\ldots, \chi(T_\beta^{-2}(x)), \chi(T_\beta^{-1}(x)), \chi(T_\beta^0(x)), \chi(T_\beta^1(x)), \chi(T_\beta^2(x)), \ldots$$

consists of alternating strings of about thirty 0's and about thirty 1's. Thus,

> if $\chi(T_\beta^{-1}(x)) = 0$ and $\chi(T_\beta^0(x)) = 1$, then
> we know that $\chi(T_\beta^k(x)) = 1$ for $k = 1, \ldots, 25$.

So the results are somewhat predictable. (In contrast, consider the fortune that could be won by predicting 25 coin tosses on a fairly regular basis!)

But that is only the beginning. The values of

$$\chi(T_\beta^{-1000}(x)), \chi(T_\beta^{-999}(x)), \chi(T_\beta^{-998}(x)), \ldots, \chi(T_\beta^{-1}(x))$$

can be used to determine the position of x fairly accurately. Using this more subtle information, one can predict far more than just 25 values of χ — it is only when $T_\beta^n(x)$ happens to land very close to 0 or $1/2$ that the value of $\chi(T_\beta^n(x))$ provides any new information. Indeed, knowing more and more values of $\chi(T_\beta^k(x))$ allows us to make longer and longer strings of predictions. In the limit, the amount of unpredictability goes to 0, so it turns out that the entropy of T_β is 0.

We remark that the relationship between entropy and unpredictability can be formalized as follows (see Exer. 2.4#12).

(2.2.5) **Proposition.** *The entropy of a transformation is 0 if and only if the past determines the future (almost surely).*

More precisely, the entropy of T is 0 if and only if, for each partition $\mathcal{A}$ of Ω into finitely many measurable sets, there is a conull subset Ω', such that, for $x, y \in \Omega'$,

if $T^k(x) \sim T^k(y)$, for all $k < 0$, then $T^k(x) \sim T^k(y)$, for all k,

where $\sim$ is the equivalence relation corresponding to the partition $\mathcal{A}$: namely, $x \sim y$ if there exists $A \in \mathcal{A}$ with $\{x, y\} \subset A$.

(2.2.6) **Example.**

1) Knowing the entire past history of a fair coin does not tell us what the next toss will be, so, in accord with Eg. 2.2.3, Prop. 2.2.5 implies the entropy of T_{Bern} is *not* 0.

2) For the Baker's Transformation, let

$$\mathcal{A} = \{[0, 1/2) \times [0, 1], [1/2, 1] \times [0, 1]\}$$

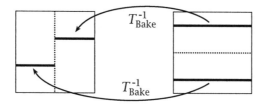

FIGURE 2.2B. The inverse image of horizontal line segments.

be the partition of the unit square into a left half and a right half. The inverse image of any horizontal line segment lies entirely in one of these halves of the square (and is horizontal) (see Fig. 2.2B), so (by induction), if x and y lie on the same horizontal line segment, then $T_{\text{Bake}}^k(x) \sim T_{\text{Bake}}^k(y)$ for all $k < 0$. On the other hand, it is (obviously) easy to find two points x and y on a horizontal line segment, such that x and y are in opposite halves of the partition. So the past does not determine the present (or the future). This is an illustration of the fact that the entropy $h(T_{\text{Bake}})$ of T_{Bake} is *not* 0 (see 2.2.2).

3) Let $\mathcal{A} = \{I, \mathbb{T} \smallsetminus I\}$, for some (proper) arc I of $\mathbb{T}$. If β is irrational, then, for any $x \in \mathbb{T}$, the set $\{ T_\beta^k(x) \mid k < 0 \}$ is dense in $\mathbb{T}$. From this observation, it is not difficult to see that if $T_\beta^k(x) \sim T_\beta^k(y)$ for all $k < 0$, then $x = y$. Hence, for an irrational rotation, the past does determine the future. This is a manifestation of the fact that the entropy $h(T_\beta)$ of T_β is 0 (see 2.2.4).

Exercise for §2.2.

#1. Show, for any x and y in the unit square, that there exists x', such that $d(x, x') < 0.1$ and $d(T_{\text{Bake}}^{20}(x), y) < 0.01$.

2.3. Definition of entropy

The fundamental difference between the behavior of the above two examples will be formalized by the notion of the entropy of a dynamical system, but, first, we define the entropy of a partition.

(2.3.1) **Remark.** Let us give some motivation for the following definition. Suppose we are interested in the location of some point ω in some probability space Ω.

- If Ω has been divided into 2 sets of equal measure, and we know which of these sets the point ω belongs to, then we have 1 bit of information about the location of ω. (The two halves of Ω can be labelled '0' and '1'.)

- If Ω has been divided into 4 sets of equal measure, and we know which of these sets the point ω belongs to, then we have 2 bits of information about the location of ω. (The four quarters of Ω can be labelled '00', '01', '10', and '11'.)

- More generally, if Ω has been divided into 2^k sets of equal measure, and we know which of these sets the point ω belongs to, then we have exactly k bits of information about the location of ω.

- The preceding observation suggests that if Ω has been divided into n sets of equal measure, and we know which of these sets the point ω belongs to, then we have exactly $\log_2 n$ bits of information about the location of ω.

- But there is no need to actually divide Ω into pieces: if we have a certain subset A of Ω, with $\mu(A) = 1/n$, and we know that ω belongs to A, then we can say that we have exactly $\log_2 n$ bits of information about the location of ω.

- More generally, if we know that ω belongs to a certain subset A of Ω, and $\mu(A) = p$, then we should say that we have exactly $\log_2(1/p)$ bits of information about the location of ω.

- Now, suppose Ω has been partitioned into finitely many subsets $A_1, \ldots, A_m$, of measure $p_1, \ldots, p_m$, and that we will be told which of these sets a random point ω belongs to. Then the right-hand side of Eq. (2.3.3) is the amount of information that we can expect (in the sense of probability theory) to receive about the location of ω.

(2.3.2) **Definition.** Suppose $\mathscr{A} = \{A_1, \ldots, A_m\}$ is a partition of a probability space (Ω, μ) into finitely many measurable sets of measure $p_1, \ldots, p_m$ respectively. (Each set A_i is called an **atom** of $\mathscr{A}$.) The **entropy** of this partition is

$$H(\mathscr{A}) = H(p_1, \ldots, p_m) = \sum_{i=1}^{m} p_i \log \frac{1}{p_i}. \qquad (2.3.3)$$

If $p_i = 0$, then $p_i \log(1/p_i) = 0$, by convention. Different authors may use different bases for the logarithm, usually either e or 2, so this definition can be varied by a scalar multiple. Note that entropy is never negative.

(2.3.4) **Remark.** Let us motivate the definition in another way. Think of the partition $\mathcal{A}$ as representing an experiment with m (mutually exclusive) possible outcomes. (The probability of the ith outcome is p_i.) We wish $H(\mathcal{A})$ to represent the amount of information one can expect to gain by performing the experiment. Alternatively, it can be thought of as the amount of uncertainty regarding the outcome of the experiment. For example, $H(\{\Omega\}) = 0$, because we gain no new information by performing an experiment whose outcome is known in advance.

Let us list some properties of H that one would expect, if it is to fit our intuitive understanding of it as the information gained from an experiment.

1) The entropy does not depend on the particular subsets chosen for the partition, but only on their probabilities. Thus, for $p_1, \ldots, p_n \geq 0$ with $\sum_i p_i = 1$, we have a real number $H(p_1, \ldots, p_n) \geq 0$. Furthermore, permuting the probabilities $p_1, \ldots, p_n$ does not change the value of the entropy $H(p_1, \ldots, p_n)$.

2) An experiment yields no information if and only if we can predict its outcome with certainty. Therefore, we have $H(p_1, \ldots, p_n) = 0$ if and only if $p_i = 1$ for some i.

3) If a certain outcome of an experiment is impossible, then there is no harm in eliminating it from the description of the experiment, so $H(p_1, \ldots, p_n, 0) = H(p_1, \ldots, p_n)$.

4) The least predictable experiment is one in which all outcomes are equally likely, so, for given n, the function $H(p_1, \ldots, p_n)$ is maximized at $(1/n, \ldots, 1/n)$.

5) $H(p_1, \ldots, p_n)$ is a continuous function of its arguments.

6) Our final property is somewhat more sophisticated, but still intuitive. Suppose we have two finite partitions $\mathcal{A}$ and $\mathcal{B}$ (not necessarily independent), and let

$$\mathcal{A} \vee \mathcal{B} = \{ A \cap B \mid A \in \mathcal{A},\ B \in \mathcal{B} \}$$

be their *join*. The join corresponds to performing **both** experiments. We would expect to get the same amount of information by performing the two experiments in succession, so

$$H(\mathcal{A} \vee \mathcal{B}) = H(\mathcal{A}) + H(\mathcal{B} \mid \mathcal{A}),$$

where $H(\mathcal{B} \mid \mathcal{A})$, the *expected* (or *conditional*) *entropy* of $\mathcal{B}$, given $\mathcal{A}$, is the amount of information expected from performing experiment $\mathcal{B}$, given that experiment $\mathcal{A}$ has already been performed.

More precisely, if experiment $\mathcal{A}$ has been performed, then some event A has been observed. The amount of information $H_A(\mathcal{B})$ now expected from experiment $\mathcal{B}$ is given by the entropy of the partition $\{B_j \cap A\}_{j=1}^n$ of A, so

$$H_A(\mathcal{B}) = H\left(\frac{\mu(B_1 \cap A)}{\mu(A)}, \frac{\mu(B_2 \cap A)}{\mu(A)}, \ldots, \frac{\mu(B_n \cap A)}{\mu(A)}\right).$$

This should be weighted by the probability of A, so

$$H(\mathcal{B} \mid \mathcal{A}) = \sum_{A \in \mathcal{A}} H_A(\mathcal{B})\,\mu(A).$$

An elementary (but certainly nontrivial) argument shows that any function H satisfying all of these conditions must be as described in Defn. 2.3.2 (for some choice of the base of the logarithm).

The entropy of a partition leads directly to the definition of the entropy of a dynamical system. To motivate this, think about repeating the same experiment every day. The first day we presumably obtain some information from the outcome of the experiment. The second day may yield some additional information (the result of an experiment — such as recording the time of sunrise — may change from day to day). And so on. If the dynamical system is "predictable," then later experiments do not yield much new information. On the other hand, in a truly unpredictable system, such as a coin toss, we learn something new every day — the expected total amount of information is directly proportional to the number of times the experiment has been repeated. The total amount of information expected to be obtained after k daily experiments (starting today) is

$$E^k(T, \mathcal{A}) = H\big(\mathcal{A} \vee T^{-1}(\mathcal{A}) \vee T^{-2}(\mathcal{A}) \vee \cdots \vee T^{-(k-1)}(\mathcal{A})\big),$$

where

$$T^\ell(\mathcal{A}) = \{\, T^\ell(A) \mid A \in \mathcal{A} \,\}$$

(see Exer. 6). It is not difficult to see that the limit

$$h(T, \mathcal{A}) = \lim_{k \to \infty} \frac{E^k(T, \mathcal{A})}{k}$$

exists (see Exer. 14). The entropy of T is the value of this limit for the most effective experiment:

(2.3.5) Definition. The *entropy* $h(T)$ is the supremum of $h(T, \mathcal{A})$ over all partitions $\mathcal{A}$ of Ω into finitely many measurable sets.

(2.3.6) **Notation.** The entropy of T may depend on the choice of the invariant measure μ, so, to avoid confusion, it may sometimes be denoted $h_\mu(T)$.

(2.3.7) **Remark.** The *entropy* of a flow is defined to be the entropy of its time-one map; that is, $h(\{\varphi_t\}) = h(\varphi_1)$.

(2.3.8) **Remark.** Although we make no use of it in these lectures, we mention that there is also a notion of *entropy* that is purely topological. Note that $E^k(T, \mathcal{A})$ is large if the partition

$$\mathcal{A} \vee T^{-1}(\mathcal{A}) \vee T^{-2}(\mathcal{A}) \vee \cdots \vee T^{-(k-1)}(\mathcal{A})$$

consists of very many small sets. In pure topology, without a measure, one cannot say whether or not the sets in a collection are "small," but one can say whether or not there are very many of them, and that is the basis of the definition. However, the topological definition uses open covers of the space, instead of measurable partitions of the space.

Specifically, suppose T is a homeomorphism of a compact metric space X.

1) For each open cover $\mathcal{A}$ of X, let $N(\mathcal{A})$ be the minimal cardinality of a subcover.

2) If $\mathcal{A}$ and $\mathcal{B}$ are open covers, let

$$\mathcal{A} \vee \mathcal{B} = \{A \cap B \mid A \in \mathcal{A},\ B \in \mathcal{B}\}.$$

3) Define

$$h_{\mathrm{top}}(T) = \sup_{\mathcal{A}} \lim_{k \to \infty} \frac{\log N(\mathcal{A} \vee T^{-1}(\mathcal{A}) \vee \cdots \vee T^{-(k-1)}(\mathcal{A}))}{k}.$$

It can be shown, for every T-invariant probability measure μ on X, that $h_\mu(T) \le h_{\mathrm{top}}(T)$.

Exercises for §2.3.

#1. Show the function H defined in Eq. (2.3.3) satisfies the formulas in:

    (a) Rem. 2.3.4(2),

    (b) Rem. 2.3.4(3),

    (c) Rem. 2.3.4(4),

    (d) Rem. 2.3.4(5), and

    (e) Rem. 2.3.4(6).

#2. Intuitively, it is clear that altering an experiment to produce more refined outcomes will not reduce the amount of information provided by the experiment. Formally, show that if $\mathcal{A} \subset \mathcal{B}$, then $H(\mathcal{A}) \le H(\mathcal{B})$. (We write $\mathcal{A} \subset \mathcal{B}$ if each atom of $\mathcal{A}$ is a union of atoms of $\mathcal{B}$ (up to measure zero).)

#3. It is easy to calculate that the entropy of a combination of experiments is precisely the sum of the entropies of the individual experiments. Intuitively, it is reasonable to expect that independent experiments provide the most information (because they have no redundancy). Formally, show

$$H(\mathcal{A}_1 \vee \cdots \vee \mathcal{A}_n) \le \sum_{i=1}^{n} H(\mathcal{A}_i).$$

[*Hint:* Assume $n = 2$ and use Lagrange Multipliers.]

#4. Show $H(\mathcal{A} \mid \mathcal{B}) \le H(\mathcal{A})$.
[*Hint:* Exer. 3.]

#5. Show $H(T^{\ell}\mathcal{A}) = H(\mathcal{A})$, for all $\ell \in \mathbb{Z}$.
[*Hint:* T^{ℓ} is measure preserving.]

#6. For $x, y \in \Omega$, show that x and y are in the same atom of $\bigvee_{\ell=0}^{k-1} T^{-\ell}(\mathcal{A})$ if and only if, for all $\ell \in \{0, \ldots, k-1\}$, the two points $T^{\ell}(x)$ and $T^{\ell}(y)$ are in the same atom of $\mathcal{A}$.

#7. Show $h(T) = h(T^{-1})$.
[*Hint:* Exer. 5 implies $E^k(T, \mathcal{A}) = E^k(T^{-1}, \mathcal{A})$.]

#8. Show $h(T^{\ell}) = |\ell| \, h(T)$, for all $\ell \in \mathbb{Z}$.
[*Hint:* For $\ell > 0$, consider $E^k(T^{\ell}, \bigvee_{i=0}^{\ell-1} T^{-i}\mathcal{A})$.]

#9. Show that entropy is an isomorphism invariant. That is, if
  - $\psi \colon (\Omega, \mu) \rightarrow (\Omega', \mu')$ is a measure-preserving map, such that
  - $\psi(T(\omega)) = T'(\psi(\omega))$ a.e.,

  then $h_{\mu}(T) = h_{\mu'}(T')$.

#10. Show $h(T, \mathcal{A}) \le H(\mathcal{A})$.
[*Hint:* Exers. 5 and 3.]

#11. Show that if $\mathcal{A} \subset \mathcal{B}$, then $h(T, \mathcal{A}) \le h(T, \mathcal{B})$.

#12. Show $|h(T, \mathcal{A}) - h(T, \mathcal{B})| \le H(\mathcal{A} \mid \mathcal{B}) + H(\mathcal{B} \mid \mathcal{A})$.
[*Hint:* Reduce to the case where $\mathcal{A} \subset \mathcal{B}$, by using the fact that $\mathcal{A} \vee \mathcal{B}$ contains both $\mathcal{A}$ and $\mathcal{B}$.]

#13. Show that the sequence $E^k(T, \mathcal{A})$ is **subadditive**; that is, $E^{k+\ell}(T, \mathcal{A}) \le E^k(T, \mathcal{A}) + E^{\ell}(T, \mathcal{A})$.

#14. Show that $\lim_{k \to \infty} \frac{1}{k} E^k(T, \mathcal{A})$ exists.
[*Hint:* Exer. 13.]

#15. Show that if T is an isometry of a compact metric space, then $h_{\text{top}}(T) = 0$.

[*Hint:* If $\mathcal{B}_\epsilon$ is the open cover by balls of radius ϵ, then $T^\ell(\mathcal{B}_\epsilon) = \mathcal{B}_\epsilon$. Choose ϵ to be a **Lebesgue number** of the open cover $\mathcal{A}$; that is, every ball of radius ϵ is contained in some element of $\mathcal{A}$.]

2.4. How to calculate entropy

Definition 2.3.5 is difficult to apply in practice, because of the supremum over all possible finite partitions. The following theorem eliminates this difficulty, by allowing us to consider only a single partition. (See Exer. 9 for the proof.)

(2.4.1) Theorem. *If $\mathcal{A}$ is any finite **generating partition** for T, that is, if*

$$\bigcup_{k=-\infty}^{\infty} T^k(\mathcal{A})$$

generates the σ-algebra of all measurable sets (up to measure 0), then

$$h(T) = h(T, \mathcal{A}).$$

(2.4.2) Corollary. *For any $\beta \in \mathbb{R}$, $h(T_\beta) = 0$.*

Proof. Let us assume β is irrational. (The other case is easy; see Exer. 1.) Let

- I be any (nonempty) proper arc of $\mathbb{T}$, and
- $\mathcal{A} = \{I, \mathbb{T} \smallsetminus I\}$.

It is easy to see that if $\mathcal{B}$ is any finite partition of $\mathbb{T}$ into connected sets, then $\#(\mathcal{A} \vee \mathcal{B}) \le 2 + \#\mathcal{B}$. Hence

$$\#\left(\mathcal{A} \vee T_\beta^{-1}(\mathcal{A}) \vee T_\beta^{-2}(\mathcal{A}) \vee \cdots \vee T_\beta^{-(k-1)}(\mathcal{A})\right) \le 2k \qquad (2.4.3)$$

(see Exer. 3), so, using 2.3.4(4), we see that

$$\begin{aligned}
E^k(T_\beta, \mathcal{A}) &= H\left(\mathcal{A} \vee T_\beta^{-1}(\mathcal{A}) \vee T_\beta^{-2}(\mathcal{A}) \vee \cdots \vee T_\beta^{-(k-1)}(\mathcal{A})\right) \\
&\le H\left(\frac{1}{2k}, \frac{1}{2k}, \ldots, \frac{1}{2k}\right) \\
&= \sum_{i=1}^{2k} \frac{1}{2k} \log(2k) \\
&= \log(2k).
\end{aligned}$$

Therefore

$$h(T_\beta, \mathcal{A}) = \lim_{k \to \infty} \frac{E^k(T_\beta, \mathcal{A})}{k} \le \lim_{k \to \infty} \frac{\log(2k)}{k} = 0.$$

One can show that $\mathcal{A}$ is a **generating partition** for T_β (see Exer. 4), so Thm. 2.4.1 implies that $h(T_\beta) = 0$. □

(2.4.4) **Corollary.** $h(T_{\text{Bern}}) = h(T_{\text{Bake}}) = \log 2$.

Proof. Because T_{Bern} and T_{Bake} are isomorphic (see 2.1.6), they have the same entropy (see Exer. 2.3#9). Thus, we need only calculate $h(T_{\text{Bern}})$.

Let
$$A = \{ f \in C^{\infty} \mid f(0) = 1 \} \qquad \text{and} \qquad \mathcal{A} = \{A, C^{\infty} \smallsetminus A\}.$$
Then
$$\mathcal{A} \vee T^{-1}(\mathcal{A}) \vee T^{-2}(\mathcal{A}) \vee \cdots \vee T^{-(k-1)}(\mathcal{A})$$
consists of the 2^k sets of the form
$$C_{\epsilon_0,\epsilon_1,\ldots,\epsilon_{k-1}} = \{ f \in C^{\infty} \mid f(\ell) = \epsilon_\ell, \text{ for } \ell = 0, 1, 2, \ldots, k-1 \},$$
each of which has measure $1/2^k$. Therefore,

$$h(T_{\text{Bern}}, \mathcal{A}) = \lim_{k \to \infty} \frac{E^k(T_{\text{Bern}}, \mathcal{A})}{k} = \lim_{k \to \infty} \frac{2^k \cdot \left[\frac{1}{2^k} \log 2^k \right]}{k}$$

$$= \lim_{k \to \infty} \frac{k \log 2}{k} = \log 2.$$

One can show that $\mathcal{A}$ is a **generating partition** for T_{Bern} (see Exer. 5), so Thm. 2.4.1 implies that $h(T_{\text{Bern}}) = \log 2$. $\qquad \square$

(2.4.5) **Remark.** One need not restrict to finite partitions; $h(T)$ is the supremum of $h(T, \mathcal{A})$, not only over all finite measurable partitions, but over all countable partitions $\mathcal{A}$, such that $H(\mathcal{A}) < \infty$. When considering countable partitions (of finite entropy), some sums have infinitely many terms, but, because the terms are positive, they can be rearranged at will. Thus, essentially the same proofs can be applied.

We noted, in Prop. 2.2.5, that if $h(T) = 0$, then the past determines the present (and the future). That is, if we know the results of all past experiments, then we can predict the result of today's experiment. Thus, 0 is the amount of information we can expect to get by performing today's experiment. More generally, Thm. 2.4.8 below shows that $h(T)$ is always the amount of information we expect to obtain by performing today's experiment.

(2.4.6) **Example.** As in Eg. 2.2.6(2), let $\mathcal{A}$ be the partition of the unit square into a left half and a right half. It is not difficult to see that

x and y lie on the same horizontal line segment

$$\Leftrightarrow \quad T_{\text{Bake}}^k(x) \sim T_{\text{Bake}}^k(y) \text{ for all } k < 0.$$

(We ignore points for which one of the coordinates is a dyadic rational — they are a set of measure zero.) Thus, the results of past experiments tell us which horizontal line segment contains the point ω (and provide no other information). The partition $\mathcal{A}$ cuts this line segment precisely in half, so the two possible results are equally likely in today's experiment; the expected amount of information is $\log 2$, which, as we know, is the entropy of T_{Bake} (see 2.4.4).

(2.4.7) Notation.

- Let $\mathcal{A}^+ = \bigvee_{\ell=1}^\infty T^\ell \mathcal{A}$. Thus, $\mathcal{A}^+$ is the partition that corresponds to knowing the results of all past experiments (see Exer. 10).

- Let $H(\mathcal{B} \mid \mathcal{A}^+)$ denote the conditional entropy of a partition $\mathcal{B}$ with respect to $\mathcal{A}^+$. More precisely:

  ◦ the measure μ has a **conditional measure** μ_A on each atom A of $\mathcal{A}^+$;

  ◦ the partition $\mathcal{B}$ induces a partition $\mathcal{B}_A$ of each atom A of $\mathcal{A}^+$;

  ◦ we have the entropy $H(\mathcal{B}_A)$ (with respect to the probability measure μ_A); and

  ◦ $H(\mathcal{B} \mid \mathcal{A}^+)$ is the integral of $H(\mathcal{B}_A)$ over all of Ω.

Thus, $H(\mathcal{B} \mid \mathcal{A}^+)$ represents the amount of information we expect to obtain by performing experiment $\mathcal{B}$, given that we know all previous results of experiment $\mathcal{A}$.

See Exer. 11 for the proof of the following theorem.

(2.4.8) Theorem. *If $\mathcal{A}$ is any finite **generating partition** for T, then $h(T) = H(\mathcal{A} \mid \mathcal{A}^+)$.*

Because $T^{-1}\mathcal{A}^+ = \mathcal{A} \vee \mathcal{A}^+$, the following corollary is immediate.

(2.4.9) Corollary. *If $\mathcal{A}$ is any finite **generating partition** for T, then $h(T) = H(T^{-1}\mathcal{A}^+ \mid \mathcal{A}^+)$.*

Exercises for §2.4.

#1. Show, directly from the definition of entropy, that if β is rational, then $h(T_\beta) = 0$.

#2. Show that if

- $\mathcal{A} = \{\mathbb{T} \smallsetminus I, I\}$, where I is a proper arc of $\mathbb{T}$, and

- $\mathcal{B}$ is a finite partition of $\mathbb{T}$ into connected sets, then

$$\sum_{C \in \mathcal{A} \vee \mathcal{B}} (\# \text{ components of } C) \le 2 + \#\mathcal{B}.$$

#3. Prove Eq. (2.4.3).

#4. Show that if
- $\mathcal{A} = \{\mathbb{T} \smallsetminus I, I\}$, where I is a nonempty, proper arc of $\mathbb{T}$, and

- β is irrational,

then $\mathcal{A}$ is a generating partition for T_β.
[*Hint:* If n is large, then $\bigvee_{k=0}^{n} T^k \mathcal{A}$ is a partition of $\mathbb{T}$ into small intervals. Thus, any open interval is a countable union of sets in $\bigcup_{k=0}^{\infty} T^k \mathcal{A}$.]

#5. Show that the partition $\mathcal{A}$ in the proof of Cor. 2.4.2 is a generating partition for T_{Bern}.

#6. The construction of a **Bernoulli shift** can be generalized, by using any probability space in place of C. Show that if a Bernoulli shift T is constructed from a measure space with probabilities $\{p_1, \ldots, p_n\}$, then $h(T) = H(p_1, \ldots, p_n)$.

#7. Show that if S is any finite, nonempty set of integers, then $h(T, \bigvee_{k \in S} T^k \mathcal{A}) = h(T, \mathcal{A})$.

#8. Suppose
- $\mathcal{A}$ is a finite **generating partition** for T,

- $\mathcal{B}$ is a finite partition of Ω, and

- $\epsilon > 0$.

Show that there is a finite set S of integers, such that $H(\mathcal{B} \mid \mathcal{A}_S) < \epsilon$, where $\mathcal{A}_S = \bigvee_{\ell \in S} T^\ell \mathcal{A}$.

#9. Show that if $\mathcal{A}$ is a finite **generating partition** for T, then $h(T) = h(T, \mathcal{A})$.
[*Hint:* Exer. 8.]

#10. For $x, y \in \Omega$, show that x and y belong to the same atom of $\mathcal{A}^+$ if and only if $T^k(x)$ and $T^k(y)$ belong to the same atom of $\mathcal{A}$, for every $k < 0$.

#11. Show:

(a) $E^k(T, \mathcal{A}) = H(\mathcal{A}) + \sum_{\ell=1}^{k-1} H(\mathcal{A} \mid \bigvee_{j=1}^{\ell} T^j \mathcal{A})$.

(b) $H(\mathcal{A} \mid \bigvee_{\ell=1}^{k} T^\ell \mathcal{A})$ is a decreasing sequence.

(c) $h(T, \mathcal{A}) = H(\mathcal{A} \mid \mathcal{A}^+)$.

(d) $h(T, \mathcal{A}) = H(T^{-1} \mathcal{A}^+ \mid \mathcal{A}^+)$.

[*Hint:* You may assume, without proof, that

$$H(\mathcal{A} \mid \bigvee_{\ell=1}^{\infty} T^{\ell}\mathcal{A}) = \lim_{k \to \infty} H(\mathcal{A} \mid \bigvee_{\ell=1}^{k} T^{\ell}\mathcal{A}),$$

if the limit exists.]

#12. Show $h(T, \mathcal{A}) = 0$ if and only if $\mathcal{A} \subset \mathcal{A}^{+}$ (up to measure zero).
[*Hint:* Exer. 11.]

2.5. Stretching and the entropy of a translation

(2.5.1) **Remark.**

1) Note that T_{β} is an isometry of $\mathbb{T}$. Hence, Cor. 2.4.2 is a particular case of the general fact that if T is an isometry (and Ω is a compact metric space), then $h(T) = 0$ (see Exers. 1 and 2.3#15).

2) Note that T_{Bake} is *far* from being an isometry of the unit square. Indeed:
   - the unit square can be foliated into horizontal line segments, and T_{Bake} stretches (local) distances on the leaves of this foliation by a factor of 2 (cf. 2.1.5), so horizontal distances grow exponentially fast (by a factor of 2^n) under iterates of T_{Bake}; and
   - distances in the complementary (vertical) direction are contracted exponentially fast (by a factor of $1/2^n$).

It is not a coincidence that $\log 2$, the logarithm of the stretching factor, is the entropy of T_{Bake}.

The following theorem states a precise relationship between stretching and entropy. (It can be stated in more general versions that apply to non-smooth maps, such as T_{Bake}.) Roughly speaking, entropy is calculated by adding contributions from all of the independent directions that are stretched at exponential rates (and ignoring directions that are contracted).

(2.5.2) **Theorem.** *Suppose*

- $\Omega = M$ *is a smooth, compact manifold,*
- *vol is a volume form on* M,
- T *is a volume-preserving diffeomorphism,*
- $\tau_1, \ldots, \tau_k$ *are (positive) real numbers, and*
- *the tangent bundle* $\mathcal{T}M$ *is a direct sum of* T-*invariant subbundles* $\mathcal{E}_1, \ldots, \mathcal{E}_k$, *such that* $\|dT(\xi)\| = \tau_i \|\xi\|$, *for each tangent vector* $\xi \in \mathcal{E}_i$,

then

$$h_{\text{vol}}(T) = \sum_{\tau_i > 1} (\dim \mathcal{E}_i) \log \tau_i.$$

(2.5.3) **Example.** If T is an isometry of M, let $\tau_1 = 1$ and $\mathcal{E}_1 = TM$. Then the theorem asserts that $h(T) = 0$. This establishes Rem. 2.5.1(1) in the special case where Ω is a smooth manifold and T is a diffeomorphism.

(2.5.4) **Example.** For T_{Bake}, let $\tau_1 = 2$, $\tau_2 = 1/2$,

$\mathcal{E}_1 = \{\text{horizontal vectors}\}$ and $\mathcal{E}_2 = \{\text{vertical vectors}\}$.

Then, if we ignore technical problems arising from the non-differentiability of T_{Bake} and the boundary of the unit square, the theorem confirms our calculation that $h(T_{\text{Bake}}) = \log 2$ (see 2.4.4).

For the special case where T is the translation by an element of G, Thm. 2.5.2 can be rephrased in the following form.

(2.5.5) **Notation.** Suppose g is an element of G

- Let $G_+ = \{ u \in G \mid \lim_{k \to -\infty} g^{-k} u g^k = e \}$. (Note that k tends to *negative* infinity.) Then G_+ is a closed, connected subgroup of G (see Exer. 5).

- Let

$$J(g, G_+) = \left| \det((\text{Ad}\, g)|_{\mathfrak{g}_+}) \right|$$

be the Jacobian of g acting by conjugation on G_+.

(2.5.6) **Remark.** G_+ is called the (expanding) *horospherical subgroup* corresponding to g. (Although this is not reflected in the notation, one should keep in mind that the subgroup G_+ depends on the choice of g.) Conjugation by g *expands* the elements of G_+ because, by definition, conjugation by g^{-1} contracts them.

There is also a *contracting* horospherical subgroup G_-, consisting of the elements that are contracted by g. It is defined by

$$G_- = \{ u \in G \mid \lim_{k \to \infty} g^{-k} u g^k = e \},$$

the only difference being that the limit is now taken as k tends to *positive* infinity. Thus, G_- is the expanding horospherical subgroup corresponding to g^{-1}.

(2.5.7) **Corollary.** *Let*

- *G be a connected Lie group,*
- *Γ be a lattice in G,*

- vol *be a G-invariant volume on* $\Gamma\backslash G$, *and*
- $g \in G$.

Then

$$h_{\text{vol}}(g) = \log J(g, G_+),$$

where, abusing notation, we write $h_{\text{vol}}(g)$ *for the entropy of the translation* $T_g \colon \Gamma\backslash G \to \Gamma\backslash G$, *defined by* $T_g(x) = xg$.

(2.5.8) **Corollary.** *If* $u \in G$ *is unipotent, then* $h_{\text{vol}}(u) = 0$.

(2.5.9) **Corollary.** *If*

- $G = \text{SL}(2, \mathbb{R})$, *and*
- $a^s = \begin{bmatrix} e^s & 0 \\ 0 & e^{-s} \end{bmatrix}$,

then $h_{\text{vol}}(a^s) = 2|s|$.

Corollary 2.5.7 calculates the entropy of g with respect to the natural volume form on $\Gamma\backslash G$. The following generalization provides an estimate (not always an exact calculation) for other invariant measures. If one accepts that entropy is determined by the amount of stretching, in the spirit of Thm. 2.5.2 and Cor. 2.5.7, then the first two parts of the following proposition are fairly obvious at an intuitive level. Namely:

(1) The hypothesis of 2.5.11(1) implies that stretching along any direction in W will contribute to the calculation of $h_\mu(a)$. This yields only an inequality, because there may be other directions, not along W, that also contribute to $h_\mu(a)$; that is, there may well be other directions that are being stretched by a and belong to the support of μ.

(2) Roughly speaking, the hypothesis of 2.5.11(2) states that any direction stretched by a and belonging to the support of μ must lie in W. Thus, only directions in W contribute to the calculation of $h_\mu(a)$. This yields only an inequality, because some directions in W may not belong to the support of μ.

(2.5.10) **Notation.** Suppose

- g is an element of G, with corresponding horospherical subgroup G_+, and
- W is a connected Lie subgroup of G_+ that is normalized by g.

Let

$$J(g, W) = \left| \det((\text{Ad } g)|_{\mathfrak{w}}) \right|$$

be the Jacobian of g on W. Thus,

$$\log J(g, W) = \sum_\lambda \log |\lambda|,$$

where the sum is over all eigenvalues of $(\operatorname{Ad} g)|_{\mathfrak{w}}$, counted with multiplicity, and $\mathfrak{w}$ is the Lie algebra of W.

(2.5.11) Proposition. *Suppose*

- *g is an element of G,*
- *μ is an measure g-invariant probability measure on $\Gamma \backslash G$, and*
- *W is a connected Lie subgroup of G_+ that is normalized by g.*

Then:

1) *If μ is W-invariant, then $h_\mu(g) \geq \log J(g, W)$.*
2) *If there is a conull, Borel subset Ω of $\Gamma \backslash G$, such that $\Omega \cap xG_- \subset xW$, for every $x \in \Omega$, then $h_\mu(g) \leq \log J(g, W)$.*
3) *If the hypotheses of (2) are satisfied, and equality holds in its conclusion, then μ is W-invariant.*

See §2.6 for a sketch of the proof.

Although we have no need for it in these lectures, let us state the following vast generalization of Thm. 2.5.2 that calculates the entropy of any diffeomorphism.

(2.5.12) Notation. Suppose T is a diffeomorphism of a smooth manifold M.

1) for each $x \in M$ and $\lambda \geq 0$, we let

$$E_\lambda(x) = \left\{ v \in \mathcal{T}_x M \ \middle| \ \limsup_{n \to \infty} \frac{\log \|d(T^{-n})_x(v)\|}{n} \leq -\lambda \right\}.$$

Note that

- $E_\lambda(x)$ is a vector subspace of $\mathcal{T}_x M$, for each x and λ, and
- we have $E_{\lambda_1}(x) \subset E_{\lambda_2}(x)$ if $\lambda_1 \leq \lambda_2$.

2) For each $\lambda > 0$, the **multiplicity** of λ at x is

$$m_x(\lambda) = \min_{\mu < \lambda}(\dim E_\lambda(x) - \dim E_\mu(x)).$$

By convention, $m_x(0) = \dim E_0(x)$.

3) We use

$$\operatorname{Lyap}(T, x) = \{\, \lambda \geq 0 \mid m_x(\lambda) \neq 0 \,\}$$

to denote the set of **Lyapunov exponents** of T at x. Note that $\sum_{\lambda \in \mathrm{Lyap}(T,x)} m_x(\lambda) = \dim M$, so $\mathrm{Lyap}(T,x)$ is a finite set, for each x.

(2.5.13) **Theorem** (Pesin's Entropy Formula). *Suppose*

- $\Omega = M$ *is a smooth, compact manifold,*
- vol *is a volume form on M, and*
- T *is a volume-preserving diffeomorphism.*

Then

$$ h_{\mathrm{vol}}(T) = \int_M \left(\sum_{\lambda \in \mathrm{Lyap}(T,x)} m_x(\lambda)\,\lambda \right) d\,\mathrm{vol}(x). $$

Exercises for §2.5.

#1. Suppose

- T is an isometry of a compact metric space Ω,
- μ is a T-invariant probability measure on Ω, and
- $\{T^{-k}x\}_{k=1}^{\infty}$ is dense in the support of μ, for a.e. $x \in \Omega$.

Use Thm. 2.4.8 (and Rem. 2.4.5) to show that $h_\mu(T) = 0$. [*Hint:* Choose a point of density x_0 for μ, and let $\mathscr{A}$ be a countable partition of Ω, such that $H(\mathscr{A}) < \infty$ and $\lim_{x \to x_0} \mathrm{diam}(\mathscr{A}(x)) = 0$, where $\mathscr{A}(x)$ denotes the atom of $\mathscr{A}$ that contains x. Show, for a conull subset of Ω, that each atom of $\mathscr{A}^+$ is a single point.]

#2. Let $G = \mathrm{SL}(2, \mathbb{R})$, and define u^t and a^s as usual (see 1.1.5). Show that if $s > 0$, then $\{u^t\}$ is the (expanding) horospherical subgroup corresponding to a^s.

#3. For each $g \in G$, show that the corresponding horospherical subgroup G_+ is indeed a **subgroup** of G.

#4. Given $g \in G$, let

$$ \mathfrak{g}_+ = \{\, v \in \mathfrak{g} \mid \lim_{k \to -\infty} v(\mathrm{Ad}_G\, g)^k = 0 \,\}. \tag{2.5.14} $$

(a) Show that $\mathfrak{g}_+$ is a Lie subalgebra of $\mathfrak{g}$.

(b) Show that if $\mathrm{Ad}_G\, g$ is diagonalizable over $\mathbb{R}$, then $\mathfrak{g}_+$ is the sum of the eigenspaces corresponding to eigenvalues of $\mathrm{Ad}_G\, g$ whose absolute value is (strictly) greater than 1.

#5. Given $g \in G$, show that the corresponding horospherical subgroup G_+ is the connected subgroup of G whose Lie algebra is $\mathfrak{g}_+$. [*Hint:* For $u \in G_+$, there is some $k \in \mathbb{Z}$ and some $v \in \mathfrak{g}$, such that $\exp v = g^{-k}ug^k$.]

#6. Derive Cor. 2.5.7 from Thm. 2.5.2, under the additional assumptions that:

(a) $\Gamma \backslash G$ is compact, and

(b) $\mathrm{Ad}_G\, g$ is diagonalizable (over $\mathbb{C}$).

#7. Derive Cor. 2.5.8 from Cor. 2.5.7.

#8. Derive Cor. 2.5.9 from Cor. 2.5.7.

#9. Give a direct proof of Cor. 2.5.8.

[*Hint:* Fix
- a small set Ω of positive measure,
- $n \colon \Omega \to \mathbb{Z}^+$ with $xu^{n(x)} \in \Omega$ for a.e. $x \in \Omega$ and $\int_\Omega n\, d\mu < \infty$,
- $\lambda > 1$, such that $d(xu^1, yu^1) < \lambda\, d(x, y)$ for $x, y \in \Gamma \backslash G$,
- a partition $\mathcal{A}$ of Ω, such that $\mathrm{diam}(\mathcal{A}(x)) \le \epsilon \lambda^{-n(x)}$ (see Exer. 10).

Use the argument of Exer. 1.]

#10. Suppose
- Ω is a precompact subset of a manifold M,
- μ is a probability measure on Ω, and
- $\rho \in L^1(\Omega, \mu)$.

Show there is a countable partition $\mathcal{A}$ of Ω, such that

(a) $\mathcal{A}$ has finite entropy, and

(b) for a.e. $x \in \Omega$, we have $\mathrm{diam}\, \mathcal{A}(x) \le e^{\rho(x)}$, where $\mathcal{A}(x)$ denotes the atom of $\mathcal{A}$ containing x.

[*Hint:* For each n, there is a partition $\mathcal{B}_n$ of Ω into sets of diameter less than $e^{-(n+1)}$, such that $\#\mathcal{B}_n \le C e^{n\dim M}$. Let $\mathcal{A}$ be the partition into the sets of the form $B_n \cap R_n$, where $B_n \in \mathcal{B}_n$ and $R_n = \rho^{-1}[n, n+1)$. Then $H(\mathcal{A}) \le H(\mathcal{R}) + \sum_{n=0}^\infty \mu(R_n) \log \#\mathcal{B}_n < \infty$.]

#11. Derive Cor. 2.5.7 from Prop. 2.5.11.

#12. Use Prop. 2.5.11 to prove

(a) Lem. 1.7.4, and

(b) Lem. 1.8.4.

#13. Derive Thm. 2.5.2 from Thm. 2.5.13.

2.6. Proof of the entropy estimate

For simplicity, we prove only the special case where g is a diagonal matrix in $G = \mathrm{SL}(2, \mathbb{R})$. The same method applies in general, if ideas from the solution of Exer. 2.5#9 are added.

(1.7.4′) **Proposition.** *Let* $G = \mathrm{SL}(2, \mathbb{R})$, *and suppose* μ *is an* a^s-*invariant probability measure on* $\Gamma \backslash G$.

1) *If* μ *is* $\{u^t\}$-*invariant, then* $h(a^s, \mu) = 2|s|$.

2) *We have* $h_\mu(a^s) \le 2|s|$.

3) *If $h_\mu(a^s) = 2|s|$ (and $s \neq 0$), then μ is $\{u^t\}$-invariant.*

(2.6.1) **Notation.**

- Let $U = \{u^t\}$.
- Let v^r be the opposite unipotent one-parameter subgroup (see 1.7.2).
- Let $a = a^s$, where $s > 0$ is sufficiently large that $e^{-s} < 1/10$, say. Note that

$$\lim_{k \to -\infty} a^{-k} u^t a^k = e \quad \text{and} \quad \lim_{k \to \infty} a^{-k} v^t a^k = e. \quad (2.6.2)$$

- Let x_0 be a point in the support of μ.
- Choose some small $\epsilon > 0$.
- Let
  - $U_\epsilon = \{ u^t \mid -\epsilon < t < \epsilon \}$, and
  - D be a small 2-disk through x_0 that is transverse to the U-orbits,

  so DU_ϵ is a neighborhood of x_0 that is naturally homeomorphic to $D \times U_\epsilon$.
- For any subset A of DU_ϵ, and any $x \in D$, the intersection $A \cap xU_\epsilon$ is called a ***plaque*** of A.

(2.6.3) **Lemma.** *There is an open neighborhood A of x_0 in DU_ϵ, such that, for any plaque F of A, and any $k \in \mathbb{Z}^+$,*

$$\text{if } F \cap Aa^k \neq \varnothing, \text{ then } F \subset Aa^k. \quad (2.6.4)$$

Proof. We may restate (2.6.4) to say:

$$\text{if } Fa^{-k} \cap A \neq \varnothing, \text{ then } Fa^{-k} \subset A.$$

Let A_0 be any very small neighborhood of x_0. If Fa^{-k} intersects A_0, then we need to add it to A. Thus, we need to add

$$A_1 = \bigcup \left\{ Fa^{-k} \;\middle|\; \begin{array}{l} F \text{ is a plaque of } A_0, \\ Fa^{-k} \cap A_0 \neq \varnothing, \\ k > 0 \end{array} \right\}.$$

This does not complete the proof, because it may be the case that, for some plaque F of A_0, a translate Fa^{-k} intersects A_1, but does not intersect A_0. Thus, we need to add more plaques to A, and continue inductively:

- Define $A_{n+1} = \bigcup \left\{ Fa^{-k} \;\middle|\; \begin{array}{l} F \text{ is a plaque of } A_0, \\ Fa^{-k} \cap A_n \neq \varnothing, \\ k > n \end{array} \right\}.$

- Let $A = \cup_{n=0}^\infty A_n$.

It is crucial to note that we may restrict to $k > n$ in the definition of A_{n+1} (see Exer. 1). Because conjugation by a^{-k} contracts distances along U exponentially (see 2.6.2), this implies that diam A is bounded by a geometric series that converges rapidly. By keeping diam A sufficiently small, we guarantee that $A \subset DU_\epsilon$. $\square$

(2.6.5) **Notation.**

- Let
$$\mathscr{A} = \{A, (\Gamma\backslash G) \smallsetminus A\},$$
where A was constructed in Lem. 2.6.3. (Technically, this is not quite correct — the proof of Lem. 2.6.7 shows that we should take a similar, but more complicated, partition of $\Gamma\backslash G$.)

- Let $\mathscr{A}^+ = \bigvee_{k=1}^{\infty} \mathscr{A}a^k$ (cf. 2.4.7).

- Let $\mathscr{A}^+(x)$ be the atom of $\mathscr{A}^+$ containing x, for each $x \in \Gamma\backslash G$.

(2.6.6) **Assumption.** Let us assume that the measure μ is measure for a. (The general case can be obtained from this by considering the ergodic decomposition of μ.)

(2.6.7) **Lemma.** *The partition $\mathscr{A}^+$ is **subordinate** to U. That is, for a.e. $x \in \Gamma\backslash G$,*

1) *$\mathscr{A}^+(x) \subset xU$, and*

2) *more precisely, $\mathscr{A}^+(x)$ is a relatively compact, open neighborhood of x (with respect to the orbit topology of xU).*

Proof. For a.e. $x \in A$, we will show that $\mathscr{A}^+(x)$ is simply the plaque of A that contains x. Thus, (1) and (2) hold for a.e. $x \in A$. By ergodicity, it immediately follows that the conditions hold for a.e. $x \in \Gamma\backslash G$.

If F is any plaque of A, then (2.6.4) implies, for each $k > 0$, that F is contained in a single atom of $\mathscr{A}a^k$. Therefore, F is contained in a single atom of $\mathscr{A}^+$. The problem is to show that $\mathscr{A}^+(x)$ contains only a single plaque.

Let $V_\epsilon = \{v^r\}_{r=-\epsilon}^{\epsilon}$, and pretend, for the moment, that $x_0 U_\epsilon V_\epsilon$ is a neighborhood of x_0. (Thus, we are we are ignoring $\{a^s\}$, and pretending that G is 2-dimensional.) For $k > 0$, we know that conjugation by a^k contracts $\{v^r\}$, so Aa^k is very thin in the $\{v^r\}$-direction (and correspondingly long in the $\{u^t\}$-direction). In the limit, we conclude that the atoms of $\mathscr{A}^+$ are infinitely thin in the $\{v^r\}$-direction. The union of any two plaques has a nonzero length in the $\{v^r\}$-direction, so we conclude that an atom of $\mathscr{A}^+$ contains only one plaque, as desired.

To complete the proof, we need to deal with the $\{a^s\}$-direction. (Unfortunately, this direction is not contracted by a^k, so the argument of the preceding paragraph does not apply.) To do this, we alter the definition of $\mathscr{A}$.

- Let $\mathscr{B}$ be a countable partition of D, such that $H(\mathscr{B}) < \infty$ and $\lim_{x \to x_0} \mathrm{diam}(\mathscr{B}(x)) = 0$.

- Let $\hat{\mathscr{A}}$ be the corresponding partition of A:
$$\hat{\mathscr{A}} = \{ (BU_\epsilon) \cap A \mid B \in \mathscr{B} \}.$$

- Let $\mathscr{A} = \hat{\mathscr{A}} \cup \{(\Gamma \backslash G) \smallsetminus A\}$.

Then $\mathscr{A}$ is a countable partition of $\Gamma \backslash G$ with $H(\mathscr{A}) < \infty$, so $h_\mu(a) = H(\mathscr{A}^+ a^{-1} \mid \mathscr{A}^+)$ (see 2.4.5 and 2.4.9). Ergodicity implies that xa^k is close to x_0 for some values of k. From the choice of $\mathscr{B}$, this implies that $\mathscr{A}^+(x)$ has small length in the $\{a^s\}$-direction. In the limit, $\mathscr{A}^+(x)$ must be infinitely thin in the $\{a^s\}$-direction. $\square$

Proof of 1.7.4′(1). We wish to show $H(\mathscr{A}^+ a^{-1} \mid \mathscr{A}^+) = 2s$ (see 2.4.9).

For any $x \in \Gamma \backslash G$, let

- μ_x be the conditional measure induced by μ on $\mathscr{A}^+(x)$, and
- λ be the Haar measure (that is, the Lebesgue measure) on xU.

Because $\mathscr{A}^+(x) \subset xU$, and μ is U-invariant, we know that

$$\mu_x \text{ is the restriction of } \lambda \text{ to } \mathscr{A}^+(x) \text{ (up to a scalar multiple).}$$
$$(2.6.8)$$

Now $\mathscr{A}^+ \subset \mathscr{A}^+ a^{-1}$, so $\mathscr{A}^+ a^{-1}$ induces a partition $\mathscr{A}_x$ of $\mathscr{A}^+(x)$. By definition, we have

$$H(\mathscr{A}^+ a^{-1} \mid \mathscr{A}^+) = -\int_{\Gamma \backslash G} \log f \, d\mu,$$

where

$$f(x) = \mu_x(\mathscr{A}_x(x)) = \frac{\lambda(\mathscr{A}_x(x))}{\lambda(\mathscr{A}^+(x))}. \qquad (2.6.9)$$

Note that, because translating by a transforms $\mathscr{A}^+ a^{-1}$ to $\mathscr{A}^+$, we have

$$x \{ u \in U \mid xu \in \mathscr{A}_x(x) \} a = xa \{ u \in U \mid xau \in \mathscr{A}^+(x) \}.$$

Conjugating by a expands λ by a factor of e^{2s}, so this implies

$$\frac{\lambda(\mathscr{A}^+(xa))}{\lambda(\mathscr{A}^+(x))} = \frac{e^{2s}\lambda(\mathscr{A}_x(x))}{\lambda(\mathscr{A}^+(x))} = e^{2s} f(x).$$

Because $0 \le f(x) \le 1$ and e^{2s} is a constant, we conclude that

$$\left(\log \lambda(\mathscr{A}^+(xa)) - \log \lambda(\mathscr{A}^+(x))\right)^+ \in L^1(\Gamma\backslash G, \mu),$$

so Lem. 2.6.10 below implies

$$-\int_{\Gamma\backslash G} \log f \, d\mu = \log e^{2s} = 2s,$$

as desired.   □

The following observation is obvious (from the invariance of μ) if $\psi \in L^1(\Gamma\backslash G, \mu)$. The general case is proved in Exer. 3.1#8.

(2.6.10) **Lemma.** *Suppose*
- *μ is an a-invariant probability measure on $\Gamma\backslash G$, and*
- *ψ is a real-valued, measurable function on $\Gamma\backslash G$,*

such that

$$(\psi(xa) - \psi(x))^+ \in L^1(\Gamma\backslash G, \mu),$$

where $(\alpha)^+ = \max(\alpha, 0)$. Then

$$\int_{\Gamma\backslash G} (\psi(xa) - \psi(x)) \, d\mu(x) = 0.$$

Proof of 1.7.4'(2). This is similar to the proof of (1). Let $\lambda_x = \lambda/\lambda(\mathscr{A}^+(x))$ be the normalization of λ to a probability measure on $\mathscr{A}^+(x)$. (In the proof of (1), we had $\lambda_x = \mu_x$ (see 2.6.8), so we did not bother to define λ_x.) Also, define

$$f_\mu(x) = \mu_x(\mathscr{A}_x(x)) \qquad \text{and} \qquad f_\lambda(x) = \lambda_x(\mathscr{A}_x(x)).$$

(In the proof of (1), we had $f_\mu = f_\lambda$ (see 2.6.9); we simply called the function f.)

We have

$$h_\mu(a) = H(\mathscr{A}^+ a^{-1} \mid \mathscr{A}^+) = -\int_{\Gamma\backslash G} \log f_\mu \, d\mu,$$

and the proof of (1) shows that

$$-\int_{\Gamma\backslash G} \log f_\lambda \, d\mu = 2s,$$

so it suffices to show

$$\int_{\Gamma\backslash G} \log f_\lambda \, d\mu \le \int_{\Gamma\backslash G} \log f_\mu \, d\mu.$$

Thus, we need only show, for a.e. $x \in \Gamma\backslash G$, that

$$\int_{\mathscr{A}^+(x)} \log f_\lambda \, d\mu_x \le \int_{\mathscr{A}^+(x)} \log f_\mu \, d\mu_x. \qquad (2.6.11)$$

Write $\mathcal{A}_x = \{A_1, \ldots, A_n\}$. For $y \in A_i$, we have

$$f_\lambda(y) = \lambda_x(A_i) \qquad \text{and} \qquad f_\mu(y) = \mu_x(A_i),$$

so

$$\int_{\mathcal{A}^+(x)} (\log f_\lambda - \log f_\mu) \, d\mu_x = \sum_{i=1}^{n} \log \frac{\lambda_x(A_i)}{\mu_x(A_i)} \mu_x(A_i).$$

Because

$$\sum_{i=1}^{n} \frac{\lambda_x(A_i)}{\mu_x(A_i)} \mu_x(A_i) = \sum_{i=1}^{n} \lambda_x(A_i) = \lambda_x(\mathcal{A}^+(x)) = 1,$$

the concavity of the log function implies

$$\sum_{i=1}^{n} \log \frac{\lambda_x(A_i)}{\mu_x(A_i)} \mu_x(A_i) \leq \log \sum_{i=1}^{n} \frac{\lambda_x(A_i)}{\mu_x(A_i)} \mu_x(A_i) \qquad (2.6.12)$$

$$= \log 1$$

$$= 0.$$

This completes the proof. $\qquad\qquad\qquad\qquad\qquad\qquad\qquad\qquad$ $\square$

Proof of 1.7.4′(3). Let μ_U be the conditional measure induced by μ on the orbit xU. To show that μ is U-invariant, we wish to show that μ is equal to λ (up to a scalar multiple).

We must have equality in the proof of 1.7.4′(2). Specifically, for a.e. $x \in \Gamma \backslash G$, we must have equality in (2.6.11), so we must have equality in (2.6.12). Because the log function is strictly concave, we conclude that

$$\frac{\lambda_x(A_i)}{\mu_x(A_i)} = \frac{\lambda_x(A_j)}{\mu_x(A_j)}$$

for all i, j. Since

$$\sum_{i=1}^{n} \lambda_x(A_i) = \lambda_x(\mathcal{A}_x) = 1 = \mu_x(\mathcal{A}_x) = \sum_{i=1}^{n} \mu_x(A_i),$$

we conclude that $\lambda_x(A_i) = \mu_x(A_i)$. This means that $\mu_U(A) = \lambda(A)$ for all atoms of $\mathcal{A}_x$. By applying the same argument with a^k in the place of a (for all $k \in \mathbb{Z}^+$), we conclude that $\mu_U(A) = \lambda(A)$ for all A in a collection that generates the σ-algebra of all measurable sets in xU. Therefore $\mu_U = \lambda$. $\qquad$ $\square$

Exercise for §2.6.

#1. In the proof of Lem. 2.6.3, show that if
- F is a plaque of A_0,
- $1 \le k \le n$, and
- $Fa^{-k} \cap A_n \ne \varnothing$,

then $Fa^{-k} \cap (A_0 \cup \cdots \cup A_{n-1}) \ne \varnothing$.
[*Hint:* Induction on k.]

Notes

The entropy of dynamical systems is a standard topic that is discussed in many textbooks, including [6, §4.3–§4.5] and [19, Chap. 4].

§2.1. Irrational rotations T_β and Bernoulli shifts T_{Bern} are standard examples. The Baker's Transformation T_{Bake} is less common, but it appears in [2, p. 22], for example.

§2.2. Proposition 2.2.5 appears in standard texts, including [19, Cor. 4.14.1].

§2.3. This material is standard (including the properties of entropy developed in the exercises).

Our treatment of the entropy of a partition is based on [7]. The elementary argument mentioned at the end of Rem. 2.3.4(6) appears in [7, pp. 9–13].

It is said that the entropy of a dynamical system was first defined by A. N. Kolmogorov [8, 9], and that much of the basic theory is due to Ya. Sinai [17, 18].

Topological entropy was defined by R. L. Adler, A. G. Konheim, and M. H. McAndrew [1]. Our discussion in Rem. 2.3.8 is taken from [5].

L. W. Goodwyn [4] proved the inequality $h_\mu(T) \le h_{\text{top}}(T)$. A simple proof of a stronger result appears in [15].

§2.4. This material is standard.
Theorem 2.4.1 is due to Ya. Sinai.

§2.5. Corollary 2.5.7 was proved by R. Bowen [3] when $\Gamma \backslash G$ is compact. The general case (when $\Gamma \backslash G$ has finite volume) was apparently already known to dynamicists in the Soviet Union. For example, it follows from the argument that proves [13, (8.35), p. 68].

A complete proof of the crucial entropy estimate (2.5.11) appears in [14, Thm. 9.7]. It is based on ideas from [11].

Pesin's Formula (2.5.13) was proved in [16]. Another proof appears in [12].

Exercise 2.5#10 is [12, Lem. 2].

§2.6. This section is based on [14, §9].

Lemma 2.6.10 is proved in [10, Prop. 2.2].

References

[1] R. L. Adler, A. G. Konheim, and M. H. McAndrew: Topological entropy. *Trans. Amer. Math. Soc.* 114 (1965), 309–319. MR 30 #5291

[2] B. Bekka and M. Mayer: *Ergodic Theory and Topological Dynamics of Group Actions on Homogeneous Spaces.* London Math. Soc. Lec. Notes #269. Cambridge U. Press, Cambridge, 2000. ISBN 0-521-66030-0, MR 2002c:37002

[3] R. Bowen: Entropy for group endomorphisms and homogeneous spaces. *Trans. Amer. Math. Soc.* 153 (1971), 401–414. MR 43 #469

[4] L. W. Goodwyn: Topological entropy bounds measure-theoretic entropy. *Proc. Amer. Math. Soc.* 23 (1969), 679–688. MR 40 #299

[5] B. Hasselblatt and A. Katok: Principal structures, in: B. Hasselblatt and A. Katok, eds., *Handbook of Dynamical Systems, Vol. 1A.* North-Holland, Amsterdam, 2002, pp. 1–203. ISBN 0-444-82669-6, MR 2004c:37001

[6] A. Katok and B. Hasselblatt: *Introduction to the Modern Theory of Dynamical Systems.* Encyclopedia of Mathematics and its Applications, 54. Cambridge Univ. Press, Cambridge, 1995. ISBN 0-521-34187-6, MR 96c:58055

[7] A. I. Khinchin: *Mathematical Foundations of Information Theory,* Dover, New York, 1957. ISBN 0-486-60434-9, MR 19,1148f

[8] A. N. Kolmogorov: A new metric invariant of transient dynamical systems and automorphisms in Lebesgue spaces (Russian). *Dokl. Akad. Nauk SSSR (N.S.)* 119 (1958), 861–864. MR 21 #2035a

[9] A. N. Kolmogorov: Entropy per unit time as a metric invariant of automorphisms (Russian). *Dokl. Akad. Nauk SSSR* 124 (1959), 754–755. MR 21 #2035b

[10] F. Ledrappier and J.-M. Strelcyn: A proof of the estimation from below in Pesin's entropy formula. *Ergodic Th. Dyn. Sys.* 2 (1982), 203–219. MR 85f:58070

[11] F. Ledrappier and L.-S. Young: The metric entropy of diffeomorphisms. I. Characterization of measures satisfying Pesin's entropy formula. *Ann. of Math.* 122 (1985), no. 3, 509–539. MR 87i:58101a

[12] R. Mañé: A proof of Pesin's formula. *Ergodic Th. Dyn. Sys.* 1 (1981), no. 1, 95–102 (errata 3 (1983), no. 1, 159–160). MR 83b:58042, MR 85f:58064

[13] G. A. Margulis: *On Some Aspects of the Theory of Anosov Systems.* Springer, Berlin, 2004. ISBN 3-540-40121-0, MR 2035655

[14] G. A. Margulis and G. M. Tomanov: Invariant measures for actions of unipotent groups over local fields on homogeneous spaces, *Invent. Math.* 116 (1994), 347–392. (Announced in *C. R. Acad. Sci. Paris Sér. I Math.* 315 (1992), no. 12, 1221–1226.) MR 94f:22016, MR 95k:22013

[15] M. Misiurewicz: A short proof of the variational principle for a $\mathbb{Z}_+^N$ action on a compact space. *Internat. Conf. on Dynam. Sys. in Math. Physics (Rennes, 1975). Astérisque* 40 (1976), 147–157. MR 56 #3250

[16] Ja. B. Pesin: Characteristic Ljapunov exponents, and smooth ergodic theory (Russian). *Uspehi Mat. Nauk* 32 (1977), no. 4 (196), 55–112, 287. Engl. transl. in *Russian Math. Surveys* 32 (1977), no. 4, 55–114. MR 57 #6667

[17] Ja. Sinaĭ: On the concept of entropy for a dynamic system (Russian). *Dokl. Akad. Nauk SSSR* 124 (1959), 768–771. MR 21 #2036a

[18] Ja. Sinaĭ: Flows with finite entropy (Russian). *Dokl. Akad. Nauk SSSR* 125 (1959), 1200–1202. MR 21 #2036b

[19] P. Walters: *An Introduction to Ergodic Theory,* Springer, New York, 1982. ISBN 0-387-90599-5, MR 84e:28017

CHAPTER 3

Facts from Ergodic Theory

This chapter simply gathers some necessary background results, mostly without proof.

3.1. Pointwise Ergodic Theorem

In the proof of Ratner's Theorem (and in many other situations), one wants to know that the orbits of a flow are uniformly distributed. It is rarely the case that *every* orbit is uniformly distributed (that is what it means to say the flow is **uniquely ergodic**), but the Pointwise Ergodic Theorem (3.1.3) shows that if the flow is "ergodic," a much weaker condition, then *almost every* orbit is uniformly distributed. (See the exercises for a proof.)

(3.1.1) **Definition.** A measure-preserving flow φ_t on a probability space (X, μ) is **ergodic** if, for each φ_t-invariant subset A of X, we have either $\mu(A) = 0$ or $\mu(A) = 1$.

(3.1.2) **Example.** For $G = \mathrm{SL}(2, \mathbb{R})$ and $\Gamma = \mathrm{SL}(2, \mathbb{Z})$, the horocycle flow η_t and the geodesic flow γ_t are ergodic on $\Gamma \backslash G$ (with respect to the Haar measure on $\Gamma \backslash G$) (see 3.2.7 and 3.2.4). These are special cases of the Moore Ergodicity Theorem (3.2.6), which implies that most flows of one-parameter subgroups on $\Gamma \backslash \mathrm{SL}(n, \mathbb{R})$ are ergodic, but the ergodicity of γ_t can easily be proved from scratch (see Exer. 3.2#3).

(3.1.3) **Theorem** (Pointwise Ergodic Theorem). *Suppose*

- *μ is a probability measure on a locally compact, separable metric space X,*
- *φ_t is an ergodic, measure-preserving flow on X, and*
- *$f \in L^1(X, \mu)$.*

Then

$$\frac{1}{T} \int_0^T f(\varphi_t(x)) \, dt \to \int_X f \, d\mu, \qquad (3.1.4)$$

for a.e. $x \in X$.

105

(3.1.5) **Definition.** A point $x \in X$ is **generic** for μ if (3.1.4) holds for every uniformly continuous, bounded function on X. In other words, a point is generic for μ if its orbit is uniformly distributed in X.

(3.1.6) **Corollary.** *If φ_t is ergodic, then almost every point of X is generic for μ.*

The converse of this corollary is true (see Exer. 2).

Exercises for §3.1.

#1. Prove Cor. 3.1.6 from Thm. 3.1.3.

#2. Let φ_t be a measure-preserving flow on (X, μ). Show that if φ_t is **not** ergodic, then almost **no** point of X is generic for μ.

#3. Let

- φ_t be an ergodic measure-preserving flow on (X, μ), and

- Ω be a non-null subset of X.

Show, for a.e. $x \in X$, that

$$\{\, t \in \mathbb{R}^+ \mid \varphi_t(x) \in \Omega \,\}$$

is unbounded.

[*Hint:* Use the Pointwise Ergodic Theorem.]

#4. Suppose

- $\phi \colon X \to X$ is a measurable bijection of X,

- μ is a ϕ-invariant probability measure on X,

- $f \in L^1(X, \mu)$, and

- $S_n(x) = f(x) + f(\phi(x)) + \cdots + f(\phi^{n-1}(x))$.

Prove the **Maximal Ergodic Theorem**: for every $\alpha \in \mathbb{R}$, if we let

$$E = \left\{\, x \in X \;\middle|\; \sup_n \frac{S_n(x)}{n} > \alpha \,\right\},$$

then $\int_E f \, d\mu \geq \alpha \mu(E)$.

[*Hint:* Assume $\alpha = 0$. Let $S_n^+(x) = \max_{0 \leq k \leq n} S_k(x)$, and $E_n = \{\, x \mid S_n^+ > 0 \,\}$, so $E = \cup_n E_n$. For $x \in E_n$, we have $f(x) \geq S_n^+(x) - S_n^+(\phi(x))$, so $\int_{E_n} f \, d\mu \geq 0$.]

#5. Prove the **Pointwise Ergodic Theorem** for ϕ, μ, f, and S_n as in Exer. 4. That is, if ϕ is ergodic, show, for a.e. x, that

$$\lim_{n \to \infty} \frac{S_n(x)}{n} = \int_X f \, d\mu.$$

[*Hint:* If $\{ x \mid \limsup S_n(x)/n > \alpha \}$ is not null, then it must be conull, by ergodicity. So the Maximal Ergodic Theorem (Exer. 4) implies $\int_X f \, d\mu \geq \alpha$.]

#6. For ϕ, μ, f, and S_n as in Exer. 4, show there is a function $f^* \in L^1(X, \mu)$, such that:

(a) for a.e. x, we have $\lim_{n \to \infty} S_n(x)/n = f^*(x)$,

(b) for a.e. x, we have $f^*(\phi(x)) = f^*(x)$, and

(c) $\int_X f^* \, d\mu = \int_X f \, d\mu$.

(This generalizes Exer. 5, because we do **not** assume ϕ is ergodic.)

[*Hint:* For $\alpha < \beta$, replacing X with the ϕ-invariant set

$$X_\alpha^\beta = \{ x \mid \liminf S_n(x)/n < \alpha < \beta < \limsup S_n(x)/n \}$$

and applying Exer. 4 yields $\int_{X_\alpha^\beta} f \, d\mu \leq \alpha \mu(X_\alpha^\beta)$ and $\int_{X_\alpha^\beta} f \, d\mu \geq \beta \mu(X_\alpha^\beta)$.]

#7. Prove Thm. 3.1.3.

[*Hint:* Assume $f \geq 0$ and apply Exer. 5 to the function $\overline{f}(x) = \int_0^1 f(\varphi_t(x)) \, dt$.]

#8. Prove Lem. 2.6.10.

[*Hint:* The Pointwise Ergodic Theorem (6) remains valid if $f = f_+ - f_-$, with $f_+ \geq 0$, $f_- \leq 0$, and $f_+ \in L^1(X, \mu)$, but the limit f^* can be $-\infty$ on a set of positive measure. Applying this to $f(x) = \psi(xa) - \psi(x)$, we conclude that

$$f^*(x) = \lim_{n \to \infty} \frac{\psi(xa^n) - \psi(x)}{n}$$

exists a.e. (but may be $-\infty$). Furthermore, $\int_{\Gamma \backslash G} f^* \, d\mu = \int_{\Gamma \backslash G} f \, dx$. Since $\psi(xa^n)/n \to 0$ in measure, there is a sequence $n_k \to \infty$, such that $\psi(xa^{n_k})/n_k \to 0$ a.e. So $f^*(x) = 0$ a.e.]

#9. Suppose

- X is a compact metric space,
- $\phi \colon X \to X$ is a homeomorphism, and
- μ is a ϕ-invariant probability measure on X.

Show ϕ is uniquely ergodic if and only if, for every continuous function f on X, there is a constant C, depending on f, such that

$$\lim_{n \to \infty} \frac{1}{n} \sum_{k=1}^{n} f(\phi^k(x)) = C,$$

uniformly over $x \in X$.

3.2. Mautner Phenomenon

We prove that the geodesic flow is ergodic (see Cor. 3.2.4). The same methods apply to many other flows on $\Gamma \backslash G$.

(3.2.1) **Definition.** Suppose φ_t is a flow on a measure space (X, μ), and f is a measurable function on X.

- f is *essentially invariant* if, for each $t \in \mathbb{R}$, we have $f(\varphi_t(x)) = f(x)$ for a.e. $x \in X$.
- f is *essentially constant* if $f(x) = f(y)$ for a.e. $x, y \in X$.

(3.2.2) **Remark.** It is obvious that any essentially constant function is essentially invariant. The converse holds if and only if φ_t is ergodic (see Exer. 1).

See Exers. 2 and 3 for the proof of the following proposition and the first corollary.

(3.2.3) **Proposition** (Mautner Phenomenon). *Suppose*

- *μ is a probability measure on $\Gamma \backslash G$,*
- *$f \in L^2(\Gamma \backslash G, \mu)$, and*
- *u^t and a^s are one-parameter subgroups of G,*

such that

- *$a^{-s} u^t a^s = u^{e^s t}$,*
- *μ is invariant under both u^t and a^s, and*
- *f is essentially a^s-invariant.*

Then f is essentially u^t-invariant.

(3.2.4) **Corollary.** *The geodesic flow y_t is ergodic on $\Gamma \backslash \mathrm{SL}(2, \mathbb{R})$.*

The following corollary is obtained by combining Prop. 3.2.3 with Rem. 3.2.2.

(3.2.5) **Corollary.** *Suppose*

- *μ is a probability measure on $\Gamma \backslash G$, and*
- *u^t and a^s are one-parameter subgroups of G,*

such that

- *$a^{-s} u^t a^s = u^{e^s t}$,*
- *μ is invariant under both u^t and a^s, and*
- *μ is ergodic for u^t.*

Then μ is ergodic for a^s.

The following result shows that flows on $\Gamma \backslash G$ are often ergodic. It is a vast generalization of the fact that the horocycle flow η_t and the geodesic flow y_t are ergodic on $\Gamma \backslash \mathrm{SL}(2, \mathbb{R})$.

(3.2.6) **Theorem** (Moore Ergodicity Theorem). *Suppose*
- *G is a connected, simple Lie group with finite center,*
- *Γ is a lattice in G, and*
- *g^t is a one-parameter subgroup of G, such that its closure $\overline{\{g^t\}}$ is not compact.*

Then g^t is ergodic on $\Gamma\backslash G$ (w.r.t. the Haar measure on $\Gamma\backslash G$).

(3.2.7) **Corollary.** *The horocycle flow η_t is ergodic on $\Gamma\backslash \mathrm{SL}(2,\mathbb{R})$.*

(3.2.8) **Remark.** The conclusion of Thm. 3.2.6 can be strengthened: not only is g^t ergodic on $\Gamma\backslash G$, but it is **mixing**. That is, if
- *A* and *B* are any two measurable subsets of $\Gamma\backslash G$, and
- μ is the *G*-invariant probability measure on $\Gamma\backslash G$,

then $(Ag^t) \cap B \to \mu(A)\,\mu(B)$ as $t \to \infty$.

The following theorem is a restatement of this remark in terms of functions.

(3.2.9) **Theorem.** *If*
- *G is a connected, simple Lie group with finite center,*
- *Γ is a lattice in G,*
- *μ is the G-invariant probability measure on $\Gamma\backslash G$, and*
- *g^t is a one-parameter subgroup of G, such that $\overline{\{g_t\}}$ is not compact,*

then
$$\lim_{t\to\infty} \int_{\Gamma\backslash G} \phi(xg^t)\,\psi(x)\,d\mu = \|\phi\|_2\,\|\psi\|_2,$$
for every $\phi, \psi \in L^2(\Gamma\backslash G, \mu)$.

(3.2.10) **Remark.** For an elementary (but very instructive) case of the following proof, assume $G = \mathrm{SL}(2,\mathbb{R})$, and $g^t = a^t$ is diagonal. Then only Case 1 is needed, and we have
$$U = \begin{bmatrix} 1 & 0 \\ * & 1 \end{bmatrix} \quad \text{and} \quad V = \begin{bmatrix} 1 & * \\ 0 & 1 \end{bmatrix}.$$
(Note that $\langle U, V \rangle = G$; that is, U and V, taken together, generate G.)

Proof. (*Requires some Lie theory and Functional Analysis*)
- Let $\mathcal{H} = 1^\perp$ be the (closed) subspace of $L^2(\Gamma\backslash G, \mu)$ consisting of the functions of integral 0. Because the desired conclusion is obvious if ϕ or ψ is constant, we may assume $\phi, \psi \in \mathcal{H}$.

- For each $g \in G$, define the unitary operator g^ρ on $\mathcal{H}$ by $(\phi g^\rho)(x) = \phi(xg^{-1})$.
- Define $\langle \phi \mid \psi \rangle = \int_{\Gamma \backslash G} \phi \psi \, d\mu$.
- Instead of taking the limit only along a one-parameter subgroup g^t, we allow a more general limit along any sequence g_j, such that $g_j \to \infty$ in G; that is, $\{g_j\}$ has no convergent subsequences.

Case 1. Assume $\{g_j\}$ is contained in a hyperbolic torus A. By passing to a subsequence, we may assume g_j^ρ converges weakly, to some operator E; that is,

$$\langle \phi g_j^\rho \mid \psi \rangle \to \langle \phi E \mid \psi \rangle \text{ for every } \phi, \psi \in \mathcal{H}.$$

Let

$$U = \{ u \in G \mid g_j u g_j^{-1} \to e \}$$

and

$$V = \{ v \in G \mid g_j^{-1} v g_j \to e \}.$$

For $v \in V$, we have

$$\begin{aligned}
\langle \phi v^\rho E \mid \psi \rangle &= \lim_{j \to \infty} \langle \phi v^\rho g_j^\rho \mid \psi \rangle \\
&= \lim_{j \to \infty} \langle \phi g_j^\rho (g_j^{-1} v g_j)^\rho \mid \psi \rangle \\
&= \lim_{j \to \infty} \langle \phi g_j^\rho \mid \psi \rangle \\
&= \langle \phi E \mid \psi \rangle,
\end{aligned}$$

so $v^\rho E = E$. Therefore, E annihilates the image of $v^\rho - I$, for every $v \in V$. Now, these images span a dense subspace of the orthogonal complement $(\mathcal{H}^V)^\perp$ of the subspace $\mathcal{H}^V$ of elements of $\mathcal{H}$ that are fixed by every element of V. Hence, E annihilates $(\mathcal{H}^V)^\perp$.

Using $*$ to denote the adjoint, we have

$$\langle \phi E^* \mid \psi \rangle = \langle \phi \mid \psi E \rangle = \lim_{j \to \infty} \langle \phi \mid \psi g_j^\rho \rangle = \lim_{j \to \infty} \langle \phi (g_j^{-1})^\rho \mid \psi \rangle,$$

so the same argument, with E^* in the place of E and g_j^{-1} in the place of g_j, shows that E^* annihilates $(\mathcal{H}^U)^\perp$.

Because g^ρ is unitary, it is normal (that is, commutes with its adjoint); thus, the limit E is also normal: we have $E^*E = EE^*$. Therefore

$$\begin{aligned}
\|\phi E\|^2 = \langle \phi E \mid \phi E \rangle &= \langle \phi(EE^*) \mid \phi \rangle \\
&= \langle \phi(E^*E) \mid \phi \rangle = \langle \phi E^* \mid \phi E^* \rangle = \|\phi E^*\|^2,
\end{aligned}$$

so $\ker E = \ker E^*$. Hence

$$\ker E = \ker E + \ker E^* \supset (\mathcal{H}^V)^\perp + (\mathcal{H}^U)^\perp$$
$$= (\mathcal{H}^V \cap \mathcal{H}^U)^\perp = (\mathcal{H}^{\langle U,V \rangle})^\perp.$$

By passing to a subsequence (so $\{g_j\}$ is contained in a single Weyl chamber), we may assume $\langle U, V \rangle = G$. Then $\mathcal{H}^{\langle U,V \rangle} = \mathcal{H}^G = 0$, so $\ker E \supset 0^\perp = \mathcal{H}$. Hence, for all $\phi, \psi \in \mathcal{H}$, we have

$$\lim \langle \phi g_j^\rho \mid \psi \rangle = \langle \phi E \mid \psi \rangle = \langle 0 \mid \psi \rangle = 0,$$

as desired.

Case 2. The general case. From the Cartan Decomposition $G = KAK$, we may write $g_j = c_j' a_j c_j$, with $c_j', c_j \in K$ and $a_j \in A$. Because K is compact, we may assume, by passing to a subsequence, that $\{c_j'\}$ and $\{c_j\}$ converge: say, $c_j' \to c'$ and $c_j \to c$. Then

$$\lim_{j \to \infty} \langle \phi g_j^\rho \mid \psi \rangle = \lim_{j \to \infty} \langle \phi(c_j' a_j c_j)^\rho \mid \psi \rangle$$
$$= \lim_{j \to \infty} \langle \phi(c_j')^\rho a_j^\rho \mid \psi(c_j^{-1})^\rho \rangle$$
$$= \lim_{j \to \infty} \langle \phi(c')^\rho a_j^\rho \mid \psi(c^{-1})^\rho \rangle$$
$$= 0,$$

by Case 1. □

(3.2.11) **Remark.**

1) If

- G and $\{g^t\}$ are as in Thm. 3.2.9,
- Γ is any discrete subgroup of G, that is **not** a lattice, and
- μ is the (infinite) G-invariant measure on $\Gamma \backslash G$,

then the above proof (with $\mathcal{H} = L^2(H \backslash G, \mu)$) shows that

$$\lim_{t \to \infty} \int_{H \backslash G} \phi(xg^t)\,\psi(x)\,d\mu = 0,$$

for every $\phi, \psi \in L^2(\Gamma \backslash G, \mu)$.

2) Furthermore, the discrete subgroup Γ can be replaced with any closed subgroup H of G, such that $H \backslash G$ has a G-invariant measure μ that is finite on compact sets.
- If the measure of $H \backslash G$ is finite, then the conclusion is as in Thm. 3.2.9.

- If the measure is infinite, then the conclusion is as in (1).

Exercises for §3.2.

#1. Suppose φ_t is a flow on a measure space (X, μ). Show that φ_t is ergodic if and only if every essentially invariant measurable function is essentially constant.

#2. Prove Prop. 3.2.3 (without quoting other theorems of the text).
[*Hint:* We have

$$f(xu) = f(xua^s) = f((xa^s)(a^{-s}ua^s)) \approx f(xa^s) = f(x),$$

because $a^{-s}ua^s \approx e$.]

#3. Derive Cor. 3.2.4 from Prop. 3.2.3.
[*Hint:* If f is essentially a^s-invariant, then the Mautner Phenomenon implies that it is also essentially u^t-invariant and essentially v^r-invariant.]

#4. Show that any mixing flow on $\Gamma \backslash G$ is ergodic.
[*Hint:* Let $A = B$ be a g^t-invariant subset of $\Gamma \backslash G$.]

#5. Derive Rem. 3.2.8 from Thm. 3.2.9.

#6. Derive Thm. 3.2.9 from Rem. 3.2.8.
[*Hint:* Any L^2 function can be approximated by step functions.]

#7. Suppose
- G and $\{g^t\}$ are as in Thm. 3.2.9,
- Γ is any discrete subgroup of G,
- μ is the G-invariant measure on $\Gamma \backslash G$,
- $\phi \in L^p(\Gamma \backslash G, \mu)$, for some $p < \infty$, and
- ϕ is essentially g^t-invariant.

Show that ϕ is essentially G-invariant.
[*Hint:* Some power of ϕ is in L^2. Use Thm. 3.2.9 and Rem. 3.2.11.]

#8. Let
- Γ be a lattice in $G = \mathrm{SL}(2, \mathbb{R})$, and
- μ be a probability measure on $\Gamma \backslash G$.

Show that if μ is invariant under both a^s and u^t, then μ is the Haar measure.
[*Hint:* Let λ be the Haar measure on $\Gamma \backslash G$, let $U_\epsilon = \{u^t \mid 0 \le t \le \epsilon\}$, and define A_ϵ and V_ϵ to be similar small intervals in $\{a^s\}$, and $\{v^r\}$, respectively. If f is continuous with compact support, then

$$\lim_{s \to \infty} \int_{yU_\epsilon A_\epsilon V_\epsilon} f(xa^s) \, d\lambda(x) = \lambda(yU_\epsilon A_\epsilon V_\epsilon) \int_{\Gamma \backslash G} f \, d\lambda,$$

for all $y \in \Gamma \backslash G$ (see Thm. 3.2.9). Because f is uniformly continuous, we see that

$$\int_{y U_\epsilon A_\epsilon V_\epsilon} f(xa^s) \, d\lambda(x) = \int_{y U_\epsilon A_\epsilon V_\epsilon a^s} f \, d\lambda$$

is approximately

$$\frac{\lambda(y U_\epsilon A_\epsilon V_\epsilon a^s)}{e^{2s\epsilon}} \int_0^{e^{2s\epsilon}} f(yu^t) \, dt.$$

By choosing y and $\{s_k\}$ such that $ya^{s_k} \to y$ and applying the Pointwise Ergodic Theorem, conclude that $\lambda = \mu$.]

3.3. Ergodic decomposition

Every measure-preserving flow can be decomposed into a union of ergodic flows.

(3.3.1) Example. Let

- $v = (\alpha, 1, 0) \in \mathbb{R}^3$, for some irrational α,
- φ_t be the corresponding flow on $\mathbb{T}^3 = \mathbb{R}^3 / \mathbb{Z}^3$, and
- μ be the Lebesgue measure on $\mathbb{T}^3$.

Then φ_t is **not** ergodic, because sets of the form $A \times \mathbb{T}^2$ are invariant.

However, the flow decomposes into a union of ergodic flows: for each $z \in \mathbb{T}$, let

- $T_z = \{z\} \times \mathbb{T}^2$, and
- μ_z be the Lebesgue measure on the torus T_z.

Then:

1) $\mathbb{T}^3$ is the disjoint union $\bigcup_z T_z$,
2) the restriction of φ_t to each subtorus T_z is ergodic (with respect to μ_z), and
3) the measure μ is the integral of the measures μ_z (by Fubini's Theorem).

The following proposition shows that every measure μ can be decomposed into ergodic measures. Each ergodic measure μ_z is called an **ergodic component** of μ.

(3.3.2) Proposition. *If μ is any φ_t-invariant probability measure on X, then there exist*

- *a measure v on a space Z, and*

- *a (measurable) family $\{\mu_z\}_{z \in Z}$ of ergodic measures on X,*

such that $\mu = \int_Z \mu_z \, dv$; that is, $\int_X f \, d\mu = \int_Z \int_X f \, d\mu_z \, dv(z)$, for every $f \in L^1(X, \mu)$.

Proof (*requires some Functional Analysis*). Let $\mathcal{M}$ be the set of φ_t-invariant probability measures on X. This is a weak*-compact, convex subset of the dual of a certain Banach space, the continuous functions on X that vanish at ∞. So Choquet's Theorem asserts that any point in $\mathcal{M}$ is a convex combination of extreme points of $\mathcal{M}$. That is, if we let Z be the set of extreme points, then there is a probability measure ν on Z, such that $\mu = \int_Z z \, d\nu(z)$. Simply letting $\mu_z = z$, and noting that the extreme points of $\mathcal{M}$ are precisely the ergodic measures (see Exer. 1) yields the desired conclusion.   □

The above proposition yields a decomposition of the measure μ, but, unlike Eg. 3.3.1, it does not provide a decomposition of the space X. However, any two ergodic measures must be mutually singular (see Exer. 2), so a little more work yields the following geometric version of the ergodic decomposition. This often allows one to reduce a general question to the case where the flow is ergodic.

(3.3.3) **Theorem** (Ergodic decomposition). *If μ is a φ_t-invariant probability measure on X, then there exist*

- *a (measurable) family $\{\mu_z\}_{z \in Z}$ of ergodic measures on X,*
- *a measure ν on Z, and*
- *a measurable function $\psi \colon X \to Z$,*

such that

1) *$\mu = \int_Z \mu_z \, d\nu$, and*
2) *μ_z is supported on $\psi^{-1}(z)$, for a.e. $z \in Z$.*

Sketch of Proof. Let $\mathcal{F} \subset L^1(X, \mu)$ be the collection of $\{0, 1\}$-valued functions that are essentially φ_t-invariant. Because the Banach space $L^1(X, \mu)$ is separable, we may choose a countable dense subset $\mathcal{F}_0 = \{\psi_n\}$ of $\mathcal{F}$. This defines a Borel function $\psi \colon X \to \{0, 1\}^\infty$. (By changing each of the functions in $\mathcal{F}_0$ on a set of measure 0, we may assume ψ is φ_t-invariant, not merely essentially invariant.) Let $Z = \{0, 1\}^\infty$ and $\nu = \psi_* \mu$. Proposition 3.3.4 below yields a (measurable) family $\{\mu_z\}_{z \in Z}$ of probability measures on X, such that (1) and (2) hold.

All that remains is to show that μ_z is ergodic for a.e. $z \in Z$. Thus, let us suppose that

$$Z_{\text{bad}} = \{ z \in Z \mid \mu_z \text{ is not ergodic} \}$$

is not a null set. For each $z \in Z_{\text{bad}}$, there is a $\{0, 1\}$-valued function $f_z \in L^1(X, \mu_z)$ that is essentially φ_t-invariant, but not essentially constant. The functions f_z can be chosen to depend measurably on z (this is a consequence of the Von Neumann Selection Theorem); thus, there is a single measurable function f on Z, such that

- $f = f_z$ a.e.$[\mu_z]$ for $z \in Z_{\text{bad}}$, and
- $f = 0$ on $Z \smallsetminus Z_{\text{bad}}$.

Because each f_z is essentially φ_t-invariant, we know that f is essentially φ_t-invariant; thus, $f \in \mathcal{F}$. On the other hand, f is not essentially constant on the fibers of ψ, so f is not in the closure of $\mathcal{F}$. This is a contradiction. $\qquad\square$

The above proof relies on the following very useful generalization of Fubini's Theorem.

(3.3.4) Proposition. *Let*

- *X and Y be complete, separable metric spaces,*
- *μ and ν be probability measures on X and Y, respectively, and*
- *$\psi: X \to Y$ be a measure-preserving map Borel map.*

Then there is a Borel map $\lambda: Y \to \text{Prob}(X)$, such that

1) *$\mu = \int_Y \lambda_y \, d\nu(y)$, and*
2) *$\lambda_y(\psi^{-1}(y)) = 1$, for all $y \in Y$.*

Furthermore, λ is unique (up to measure zero).

Exercises for §3.3.

#1. In the notation of the proof of Prop. 3.3.2, show that a point μ of $\mathcal{M}$ is ergodic if and only if it is an extreme point of $\mathcal{M}$. (A point μ of $\mathcal{M}$ is an **extreme point** if it is **not** a convex combination of two other points of $\mathcal{M}$; that is, if there do not exist $\mu_1, \mu_2 \in \mathcal{M}$, and $t \in (0,1)$, such that $\mu = t\mu_1 + (1-t)\mu_2$ and $\mu_1 \neq \mu_2$.)

#2. Suppose μ_1 and μ_2 are ergodic, φ_t-invariant probability measures on X. Show that if $\mu_1 \neq \mu_2$, then there exist subsets Ω_1 and Ω_2 of X, such that, for $i, j \in \{1, 2\}$, we have

$$\mu_i(\Omega_j) = \begin{cases} 1 & \text{if } i = j, \\ 0 & \text{if } i \neq j. \end{cases}$$

#3. Let

- *X and X' be complete, separable metric spaces,*
- *μ and μ' be probability measures on X and X', respectively,*
- *φ_t and φ'_t be ergodic, measure-preserving flows on X and X', respectively,*

- $\psi \colon X \to Y$ be a measure-preserving map, equivariant Borel map, and

- Ω be a conull subset of X, such that $\psi^{-1}(y) \cap \Omega$ is countable, for a.e. $y \in Y$.

Show there is a conull subset Ω' of X, such that $\psi^{-1}(y) \cap \Omega$ is finite, for a.e. $y \in Y$.

[*Hint:* The function $f(x) = \lambda_{\psi(x)}(\{x\})$ is essentially φ_t-invariant, so it must be essentially constant. A probability measure with all atoms of the same weight must have only finitely many atoms.]

3.4. Averaging sets

The proof of Ratner's Theorem uses a version of the Pointwise Ergodic Theorem that applies to (unipotent) groups that are not just one-dimensional. The classical version (3.1.3) asserts that averaging a function over larger and larger intervals of almost any orbit will converge to the integral of the function. Note that the average is over intervals, not over arbitrary large subsets of the orbit. In the setting of higher-dimensional groups, we will average over "averaging sets."

(3.4.1) **Definition.** Suppose

- U is a connected, unipotent subgroup of G,
- a is a hyperbolic element of G that normalizes U
- $a^{-n}ua^n \to e$ as $n \to -\infty$ (note that this is $-\infty$, not ∞!), and
- E is a ball in U (or, more generally, E is any bounded, non-null, Borel subset of U).

Then:

1) we say that a is an **expanding automorphism** of U,

2) for each $n \geq 0$, we call $E_n = a^{-n}Ea^n$ an **averaging set**, and

3) we call $\{E_n\}_{n=0}^{\infty}$ an **averaging sequence**.

(3.4.2) **Remark.**

1) By assumption, conjugating by a^n contracts U when $n < 0$. Conversely, conjugating by a^n expands U when $n > 0$. Thus, $E_1, E_2, \ldots$ are larger and larger subsets of U. (This justifies calling a an "expanding" automorphism.)

2) Typically, one takes E to be a nice set (perhaps a ball) that contains e, with $E \subset a^{-1}Ea$. In this case, $\{E_n\}_{n=0}^{\infty}$ is an increasing Følner sequence (see Exer. 1), but, for technical reasons, we will employ a more general choice of E at

one point in our argument (namely, in 5.8.7, the proof of Prop. 5.2.4′).

(3.4.3) Theorem (Pointwise Ergodic Theorem). *If*

- *U is a connected, unipotent subgroup of G,*
- *a is an **expanding automorphism** of U,*
- *ν_U is the Haar measure on U,*
- *μ is an ergodic U-invariant probability measure on $\Gamma\backslash G$, and*
- *f is a continuous function on $\Gamma\backslash G$ with compact support,*

then there exists a U-invariant subset Ω of $\Gamma\backslash G$ with $\mu(\Omega) = 1$, such that

$$\frac{1}{\nu_U(E_n)} \int_{E_n} f(xu)\, d\nu_U(u) \to \int_{\Gamma\backslash G} f(y)\, d\mu(y) \quad \text{as } n \to \infty.$$

for every $x \in \Omega$ and every averaging sequence $\{E_n\}$ in U.

To overcome some technical difficulties, we will also use the following uniform approximate version (see Exer. 3). It is "uniform," because the same number N works for all points $x \in \Omega_\epsilon$, and the same set Ω_ϵ works for all functions f.

(3.4.4) Corollary (Uniform Pointwise Ergodic Theorem). *If*

- *U is a connected, unipotent subgroup of G,*
- *a is an **expanding automorphism** of U,*
- *ν_U is the Haar measure on U,*
- *μ is an ergodic U-invariant probability measure on $\Gamma\backslash G$, and*
- *$\epsilon > 0$,*

then there exists a subset Ω_ϵ of $\Gamma\backslash G$ with $\mu(\Omega_\epsilon) > 1 - \epsilon$, such that for

- *every continuous function f on $\Gamma\backslash G$ with compact support,*
- *every averaging sequence $\{E_n\}$ in U, and*
- *every $\delta > 0$,*

there is some $N \in \mathbb{N}$, such that

$$\left| \frac{1}{\nu_U(E_n)} \int_{E_n} f(xu)\, d\nu_U(u) - \int_{\Gamma\backslash G} f(y)\, d\mu(y) \right| < \delta,$$

for all $x \in \Omega_\epsilon$ and all $n \geq N$.

(3.4.5) Remark.

1) A Lie group G said to be **amenable** if it has a Følner sequence.

2) It is known that a **connected** Lie group G is amenable if and only if there are closed, connected, normal subgroups U and R of G, such that
   - U is unipotent,
   - $U \subset R$,
   - R/U is abelian, and
   - G/R is compact.

3) There are examples to show that not every Følner sequence $\{E_n\}$ can be used as an averaging sequence, but it is always the case that some subsequence of $\{E_n\}$ can be used as the averaging sequence for a pointwise ergodic theorem.

Exercises for §3.4.

#1. Suppose
   - U is a connected, unipotent subgroup of G,
   - a is an **expanding automorphism** of U,
   - ν_U is the Haar measure on U, and
   - E is a precompact, open subset of U, such that $a^{-1}Ea \subset E$.

Show that the averaging sequence E_n is an increasing **Følner sequence**; that is,

(a) for each nonempty compact subset C of U, we have $\nu_U((CE_n) \triangle E_n)/\nu_U(E_n) \to 0$ as $n \to \infty$, and

(b) $E_n \subset E_{n+1}$, for each n.

#2. Show that if G is amenable, then there is an invariant probability measure for any action of G on a compact metric space. More precisely, suppose
   - $\{E_n\}$ is a Følner sequence in a Lie group G,
   - X is a compact metric space, and
   - G acts continuously on X.

Show there is a G-invariant probability measure on X.
[*Hint:* Haar measure restricts to a measure ν_n on E_n. Pushing this to X (and normalizing) yields a probability measure μ_n on X. Any weak*-limit of $\{\mu_n\}$ is G-invariant.]

#3. Derive Cor. 3.4.4 from Thm. 3.4.3.

Notes

A few of the many introductory books on Ergodic Theory are [7, 8, 23].

§3.1. This material is standard.

The Pointwise Ergodic Theorem is due to G. D. Birkhoff [2]. There are now many different proofs, such as [9, 10]. (See also [1, Thm. I.2.5, p. 17]). The hints for Exers. 3.1#4 and 3.1#5 are adapted from [5, pp. 19-24].

Exercise 3.1#8 is [11, Prop. 2.2].

A solution to Exer. 3.1#9 appears in [1, Thm. I.3.8, p. 33].

§3.2. The Moore Ergodicity Theorem (3.2.6) was first proved by C. C. Moore [16]. Later, he [17] extended this to a very general version of the Mautner Phenomenon (3.2.3).

Mixing is a standard topic (see, e.g., [8, 23] and [1, pp. 21-28].) Our proof of Thm. 3.2.9 is taken from [3]. Proofs can also be found in [1, Chap. 3], [13, §II.3] and [24, Chap. 2].

A solution to Exer. 3.2#1 appears in [1, Thm. I.1.3, p. 3].

The hint to Exer. 3.2#8 is adapted from [14, Lem. 5.2, p. 31].

§3.3. This material is standard.

A complete proof of Prop. 3.3.2 from Choquet's Theorem appears in [19, §12].

See [18, §8] for a brief history (and proof) of the ergodic decomposition (3.3.3).

Proposition 3.3.4 appears in [20, §3].

Exer. 3.3#1 is solved in [1, Prop. 3.1, p. 30].

§3.4. For any amenable Lie group, a theorem of A. Tempelman [21], generalized by W. R. Emerson [4], states that certain Følner sequences can be used as averaging sequences in a pointwise ergodic theorem. (A proof also appears in [22, Cor. 6.3.2, p. 218].) The Uniform Pointwise Ergodic Theorem (3.4.4) is deduced from this in [15, §7.2 and §7.3].

The book of Greenleaf [6] is the classic source for information on amenable groups.

The converse of Exer. 3.4#2 is true [6, Thm. 3.6.2]. Indeed, the existence of invariant measures is often taken as the definition of amenability. See [24, §4.1] for a discussion of amenable groups from this point of view, including the characterization mentioned in Rem. 3.4.5(2).

Remark 3.4.5(3) is a theorem of E. Lindenstrauss [12].

References

[1] B. Bekka and M. Mayer: *Ergodic Theory and Topological Dynamics of Group Actions on Homogeneous Spaces.* London Math. Soc. Lec. Notes #269. Cambridge U. Press, Cambridge, 2000. ISBN 0-521-66030-0, MR 2002c:37002

[2] G. D. Birkhoff: Proof of the ergodic theorem. *Proc. Natl. Acad. Sci. USA* 17 (1931), 656-660. Zbl 0003.25602

[3] R. Ellis and M. Nerurkar: Enveloping semigroup in ergodic theory and a proof of Moore's ergodicity theorem, in: J. C. Alexander., ed., *Dynamical Systems (College Park, MD, 1986-87)*, 172-179, Lecture Notes in Math. #1342, Springer, Berlin-New York, 1988. ISBN 3-540-50174-6, MR 90g:28024

[4] W. R. Emerson: The pointwise ergodic theorem for amenable groups. *Amer. J. Math.* 96 (1974), no. 3, 472-478. MR 50 #7403

[5] N. A. Friedman: *Introduction to Ergodic Theory.* Van Nostrand, New York, 1970. MR 55 #8310

[6] F. P. Greenleaf: *Invariant Means on Topological Groups.* Van Nostrand, New York, 1969. MR 40 #4776, MR0251549

[7] P. R. Halmos: *Lectures on Ergodic Theory.* Chelsea, New York, 1956. Zbl 0096.09004

[8] A. Katok and B. Hasselblatt: *Introduction to the Modern Theory of Dynamical Systems.* Cambridge University Press, Cambridge, 1995. ISBN 0-521-34187-6, MR 96c:58055

[9] T. Kamae: A simple proof of the ergodic theorem using nonstandard analysis. *Israel J. Math.* 42 (1982), no. 4, 284-290. MR 84i:28019

[10] Y. Katznelson and B. Weiss: A simple proof of some ergodic theorems. *Israel J. Math.* 42 (1982), no. 4, 291-296. MR 84i:28020

[11] F. Ledrappier and J.-M. Strelcyn A proof of the estimation from below in Pesin's entropy formula. *Ergodic Theory Dynam. Systems* 2 (1982), no. 2, 203-219. MR 85f:58070

[12] E. Lindenstrauss: Pointwise theorems for amenable groups. *Invent. Math.* 146 (2001), no. 2, 259-295. MR 2002h:37005

[13] G.A. Margulis: *Discrete Subgroups of Semisimple Lie Groups.* Springer, Berlin, 1991 ISBN 3-540-12179-X, MR 92h:22021

[14] G. A. Margulis: *On Some Aspects of the Theory of Anosov Systems.* Springer, Berlin, 2004. ISBN 3-540-40121-0, MR 2035655

[15] G. A. Margulis and G. M. Tomanov: Invariant measures for actions of unipotent groups over local fields on homogeneous spaces. *Invent. Math.* 116 (1994), 347–392. MR 95k:22013

[16] C. C. Moore: Ergodicity of flows on homogeneous spaces. *Amer. J. Math.* 88 1966 154–178. MR 33 #1409

[17] C. C. Moore: The Mautner phenomenon for general unitary representations. *Pacific J. Math.* 86 (1980), 155–169. MR 81k:22010

[18] J. C. Oxtoby: Ergodic sets. *Bull. Amer. Math. Soc.* 58 (1952), 116–136. MR 13,850e

[19] R. R. Phelps: *Lectures on Choquet's theorem, 2nd ed.* Lecture Notes in Mathematics #1757. Springer-Verlag, Berlin, 2001. ISBN 3-540-41834-2, MR 2002k:46001

[20] V. A. Rohlin: On the fundamental ideas of measure theory (Russian). *Mat. Sbornik N.S.* 25(67), (1949), 107–150. (English transl. in: *Amer. Math. Soc. Transl.* (1) 10 (1962), 1–54.) MR 13,924e

[21] A. Tempelman: Ergodic theorems for general dynamic systems. *Dokl. Akad. Nauk SSSR,* 176 (1967), no. 4, 790–793. (English translation: *Soviet Math. Dokl.,* 8 (1967), no. 5, 1213–1216.) MR 36 #2779

[22] A. Tempelman: *Ergodic Theorems for Group Actions. Informational and Thermodynamical Aspects.* (Translated and revised from the 1986 Russian original. Mathematics and its Applications, 78.) Kluwer, Dordrecht, 1992. ISBN 0-7923-1717-3, MR 94f:22007

[23] P. Walters: *An Introduction to Ergodic Theory.* Springer, New York, 1982 ISBN 0-387-90599-5, MR 84e:28017

[24] R. J. Zimmer: *Ergodic Theory and Semisimple Groups.* Birkhäuser, Boston, 1984. ISBN 3-7643-3184-4, MR 86j:22014

CHAPTER 4

Facts about Algebraic Groups

In the theory of Lie groups, all homomorphisms (and other maps) are generally assumed to be C^∞ functions (see §4.9). The theory of algebraic groups describes the conclusions that can be obtained from the stronger assumption that the maps are polynomial functions (or, at least, rational functions). Because the polynomial nature of unipotent flows plays such an important role in the arguments of Chapter 1 (see, for example, Prop. 1.5.15), it is natural to expect that a good understanding of polynomials will be essential at some points in the more complete proof presented in Chapter 5. However, the reader may wish to skip over this chapter, and refer back when necessary.

4.1. Algebraic groups

(4.1.1) Definition.

- We use $\mathbb{R}[x_{1,1}, \ldots, x_{\ell,\ell}]$ to denote the set of real polynomials in the ℓ^2 variables $\{x_{i,j} \mid 1 \le i, j \le \ell\}$.

- For any $Q \in \mathbb{R}[x_{1,1}, \ldots, x_{\ell,\ell}]$, and any $n \times n$ matrix g, we use $Q(g)$ to denote the value obtained by substituting the matrix entries $g_{i,j}$ into the variables $x_{i,j}$. For example:
  - If $Q = x_{1,1} + x_{2,2} + \cdots + x_{\ell,\ell}$, then $Q(g)$ is the trace of g.
  - If $Q = x_{1,1}x_{2,2} - x_{1,2}x_{2,1}$, then $Q(g)$ is the determinant of the first principal 2×2 minor of g.

- For any subset $\mathcal{Q}$ of $\mathbb{R}[x_{1,1}, \ldots, x_{\ell,\ell}]$, let

  $$\mathrm{Var}(\mathcal{Q}) = \{g \in \mathrm{SL}(\ell, \mathbb{R}) \mid Q(g) = 0, \ \forall Q \in \mathcal{Q}\}.$$

  This is the **variety** associated to $\mathcal{Q}$.

- A subset H of $\mathrm{SL}(\ell, \mathbb{R})$ is **Zariski closed** if there is a subset $\mathcal{Q}$ of $\mathbb{R}[x_{1,1}, \ldots, x_{\ell,\ell}]$, such that $H = \mathrm{Var}(\mathcal{Q})$. (In the special case where H is a sub*group* of $\mathrm{SL}(\ell, \mathbb{R})$, we may also say that H is a **real algebraic group** or an **algebraic group that is defined over** $\mathbb{R}$.)

(4.1.2) Example. Each of the following is a real algebraic group (see Exer. 2):

1) $SL(\ell, \mathbb{R})$.

2) The group

$$\mathbb{D}_\ell = \begin{bmatrix} * & & 0 \\ & \ddots & \\ 0 & & * \end{bmatrix} \subset SL(\ell, \mathbb{R})$$

of diagonal matrices in $SL(\ell, \mathbb{R})$.

3) The group

$$\mathbb{U}_\ell = \begin{bmatrix} 1 & & 0 \\ & \ddots & \\ * & & 1 \end{bmatrix} \subset SL(\ell, \mathbb{R})$$

of lower-triangular matrices with 1's on the diagonal.

4) The group

$$\mathbb{D}_\ell \mathbb{U}_\ell = \begin{bmatrix} * & & 0 \\ & \ddots & \\ * & & * \end{bmatrix} \subset SL(\ell, \mathbb{R})$$

of lower-triangular matrices in $SL(\ell, \mathbb{R})$.

5) The copy of $SL(n, \mathbb{R})$ in the top left corner of $SL(\ell, \mathbb{R})$ (if $n < \ell$).

6) The *stabilizer*

$$\text{Stab}_{SL(\ell,\mathbb{R})}(v) = \{\, g \in SL(\ell, \mathbb{R}) \mid vg = v \,\}$$

of any vector $v \in \mathbb{R}^\ell$.

7) The *stabilizer*

$$\text{Stab}_{SL(\ell,\mathbb{R})}(V) = \{\, g \in SL(\ell, \mathbb{R}) \mid \forall v \in V,\ vg \in V \,\}$$

of any linear subspace V of $\mathbb{R}^\ell$.

8) The special orthogonal group $SO(Q)$ of a quadratic form Q on $\mathbb{R}^\ell$ (see Defn. 1.2.1).

It is important to realize that most closed subsets of $SL(\ell, \mathbb{R})$ are *not* Zariski closed. In particular, the following important theorem tells us that an infinite, discrete subset can never be Zariski closed. (It is a generalization of the fact that any nontrivial polynomial function on $\mathbb{R}$ has only finitely many zeroes.) We omit the proof.

(4.1.3) **Theorem** (Whitney). *Any Zariski closed subset of* $\mathrm{SL}(\ell, \mathbb{R})$ *has only finitely many components (with respect to the usual topology of* $\mathrm{SL}(\ell, \mathbb{R})$ *as a Lie group).*

(4.1.4) **Example.** From Thm. 4.1.3, we know that the discrete group $\mathrm{SL}(\ell, \mathbb{Z})$ is **not** Zariski closed. In fact, we will see that $\mathrm{SL}(\ell, \mathbb{Z})$ is not contained in any Zariski closed, proper subgroup of $\mathrm{SL}(\ell, \mathbb{R})$ (see Exer. 4.7#1).

(4.1.5) **Remark.** Zariski closed sets need not be submanifolds of $\mathrm{SL}(\ell, \mathbb{R})$. This follows from Exer. 4, for example, because the union of two submanifolds that intersect is usually not a submanifold — the intersection is a singularity.

Exercise 10 defines the **dimension** of any Zariski closed set Z. Although we do not prove this, it can be shown that (if Z is nonempty), there is a unique smallest Zariski closed subset S of Z, such that

- $\dim S < \dim Z$,
- $Z \smallsetminus S$ is a C^∞ submanifold of $\mathrm{SL}(\ell, \mathbb{R})$, and
- $\dim Z$ (as defined below) is equal to the dimension of $Z \smallsetminus S$ as a manifold.

The set S is the **singular set** of Z. From the uniqueness of S, it follows that any Zariski closed sub**group** of $\mathrm{SL}(\ell, \mathbb{R})$ is a C^∞ submanifold of $\mathrm{SL}(\ell, \mathbb{R})$ (see Exer. 5);

Exercises for §4.1.

#1. Show that every Zariski closed subset of $\mathrm{SL}(\ell, \mathbb{R})$ is closed (in the usual topology of $\mathrm{SL}(\ell, \mathbb{R})$ as a Lie group).

#2. Verify that each of the groups in Eg. 4.1.2 is Zariski closed. [*Hint:* (1) Let $Q = \varnothing$. (2) Let $Q = \{x_{i,j} \mid i \neq j\}$. (6) Let $Q = \{v_1 x_{1,j} + \cdots + v_\ell x_{\ell,j} - v_j \mid 1 \leq j \leq \ell\}$, where $v = (v_1, \ldots, v_\ell)$.]

#3. Show that if Z is a Zariski closed subset of $\mathrm{SL}(\ell, \mathbb{R})$, and $g \in \mathrm{SL}(\ell, \mathbb{R})$, then Zg is Zariski closed.

#4. Suppose Z_1 and Z_2 are Zariski closed subsets of $\mathrm{SL}(\ell, \mathbb{R})$. Show that the union $Z_1 \cup Z_2$ is Zariski closed.

#5. Show that if G is a Zariski closed subgroup of $\mathrm{SL}(\ell, \mathbb{R})$, then G is a C^∞ submanifold of $\mathrm{SL}(\ell, \mathbb{R})$ (so G is a Lie group). [*Hint:* Uniqueness of the singular set S (see 4.1.5) implies $Sg = S$ for all $g \in G$, so $S = \varnothing$.]

The remaining exercises present some (more technical) information about Zariski closed sets, including the notion of dimension.

#6. For any subset Z of $SL(\ell, \mathbb{R})$, let $\mathscr{I}(Z)$ be the collection of polynomials that vanish on Z; that is,

$$\mathscr{I}(Z) = \{\, Q \in \mathbb{R}[x_{1,1}, \ldots, x_{\ell,\ell}] \mid \forall z \in Z, \ Q(z) = 0 \,\}.$$

(a) Show Z is Zariski closed if and only if $Z = \mathrm{Var}(\mathscr{I}(Z))$.

(b) Show that $\mathscr{I}(Z)$ is an ideal; that is,

    (i) $0 \in \mathscr{I}(Z)$,

    (ii) for all $Q_1, Q_2 \in \mathscr{I}(Z)$, we have $Q_1 + Q_2 \in \mathscr{I}(Z)$, and

    (iii) for all $Q_1 \in \mathscr{I}(Z)$ and $Q_2 \in \mathbb{R}[x_{1,1}, \ldots, x_{\ell,\ell}]$, we have $Q_1 Q_2 \in \mathscr{I}(Z)$.

#7. Recall that a ring R is **Noetherian** if it has the ascending chain condition on ideas; this means that if $I_1 \subset I_2 \subset \cdots$ is any increasing chain of ideals, then we have $I_n = I_{n+1} = \cdots$ for some n.

(a) Show that a commutative ring R is Noetherian if and only if all of its ideals are finitely generated; that is, for each ideal I of R, there is a **finite** subset F of I, such that I is the smallest ideal of R that contains F.

(b) Show that $\mathbb{R}[x_{1,1}, \ldots, x_{\ell,\ell}]$ is Noetherian.

(c) Show that if Z is a Zariski closed subset of $SL(\ell, \mathbb{R})$, then there is a **finite** subset Q of $\mathbb{R}[x_{1,1}, \ldots, x_{\ell,\ell}]$, such that $Z = \mathrm{Var}(Q)$.

(d) Prove that the collection of Zariski closed subsets of $SL(\ell, \mathbb{R})$ has the descending chain condition: if $Z_1 \supset Z_2 \supset \cdots$ is a decreasing chain of Zariski closed sets, then we have $Z_n = Z_{n+1} = \cdots$ for some n.

[*Hint:* (7b) Show that if R is Noetherian, then the polynomial ring $R[x]$ is Noetherian: If I is an ideal in $R[x]$, let

$$I_n = \{\, r \in R \mid \exists Q \in R[x], \ rx^n + Q \in I \text{ and } \deg Q < n \,\}.$$

Then $I_n \subset I_{n+1} \subset \cdots$ is an increasing chain of ideals.]

#8. A Zariski closed subset of $SL(\ell, \mathbb{R})$ is **irreducible** if it is **not** the union of two Zariski closed **proper** subsets.

Let Z be a Zariski closed subset of $SL(\ell, \mathbb{R})$.

(a) Show that Z is the union of finitely many irreducible Zariski closed subsets.

(b) An **irreducible component** of Z is an irreducible Zariski closed subset of Z that is not not properly contained in any irreducible Zariski closed subset of Z.

    (i) Show that Z is the union of its irreducible components.

(ii) Show that Z has only finitely many irreducible components.

[*Hint:* (8a) Proof by contradiction: use the descending chain condition. (8b) Use (8a).]

#9. Suppose G is a Zariski closed subgroup of $\mathrm{SL}(\ell, \mathbb{R})$.

(a) Show that the irreducible components of G are disjoint.

(b) Show that the irreducible components of G are cosets of a Zariski closed subgroup of G.

#10. The **dimension** of a Zariski closed set Z is the largest r, such that there is a chain $Z_0 \subset Z_1 \subset \cdots \subset Z_r$ of nonempty, irreducible Zariski closed subsets of Z.

It can be shown (and you may assume) that dim Z is the largest r, for which there is a linear map $T \colon \mathrm{Mat}_{\ell \times \ell}(\mathbb{R}) \to \mathbb{R}^r$, such that $T(Z)$ contains a nonempty open subset of $\mathbb{R}^r$.

(a) Show dim $Z = 0$ if and only if Z is finite and nonempty.

(b) Show dim $Z_1 \le \dim Z_2$ if $Z_1 \subset Z_2$.

(c) Show $\dim(Z_1 \cup Z_2) = \max\{\dim Z_1, \dim Z_2\}$ if Z_1 and Z_2 are Zariski closed.

(d) Show $\dim \mathrm{SL}(\ell, \mathbb{R}) = \ell^2 - 1$.

(e) Show that the collection of irreducible Zariski closed subsets of $\mathrm{SL}(\ell, \mathbb{R})$ has the ascending chain condition: if $Z_1 \subset Z_2 \subset \cdots$ is an increasing chain of irreducible Zariski closed sets, then we have $Z_n = Z_{n+1} = \cdots$ for some n.

#11. Suppose V and W are Zariski closed sets in $\mathrm{SL}(\ell, \mathbb{R})$. Show that if

- $V \subset W$,
- W is irreducible, and
- $\dim V = \dim W$,

then $V = W$.

4.2. Zariski closure

(4.2.1) **Definition.** The **Zariski closure** of a subset H of $\mathrm{SL}(\ell, \mathbb{R})$ is the (unique) smallest Zariski closed subset of $\mathrm{SL}(\ell, \mathbb{R})$ that contains H (see Exer. 1). We use $\overline{\overline{H}}$ to denote the Zariski closure of H.

(4.2.2) **Remark.**

1) Obviously, H is Zariski closed if and only if $\overline{\overline{H}} = H$.

2) One can show that if H is a subgroup of $\mathrm{SL}(\ell, \mathbb{R})$, then $\overline{\overline{H}}$ is also a subgroup of $\mathrm{SL}(\ell, \mathbb{R})$ (see Exer. 4.3#11).

Every Zariski closed subgroup of $\mathrm{SL}(\ell, \mathbb{R})$ is closed (see Exer. 4.1#1) and has only finitely many connected components (see 4.1.3). The converse is false:

(4.2.3) **Example.** Let

$$A = \left\{ \begin{bmatrix} t & 0 \\ 0 & 1/t \end{bmatrix} \,\middle|\, t \in \mathbb{R}^+ \right\} \subset \mathrm{SL}(2, \mathbb{R}).$$

Then
1) A is closed,
2) A is connected (so it has only one connected component), and
3) $\overline{\overline{A}} = \left\{ \begin{bmatrix} t & 0 \\ 0 & 1/t \end{bmatrix} \,\middle|\, t \in \mathbb{R} \smallsetminus \{0\} \right\}$ (see Exer. 3).

So $\overline{\overline{A}} \cong \mathbb{R} \smallsetminus \{0\}$ has two connected components. Since $\overline{\overline{A}} \neq A$, we know that A is not Zariski closed.

Although A is not exactly equal to $\overline{\overline{A}}$ in Eg. 4.2.3, there is very little difference: A has finite index in $\overline{\overline{A}}$. For most purposes, a finite group can be ignored, so we make the following definition.

(4.2.4) **Definition.** A subgroup H of $\mathrm{SL}(\ell, \mathbb{R})$ is *almost Zariski closed* if H is a finite-index subgroup of $\overline{\overline{H}}$.

(4.2.5) **Remark.** Any finite-index subgroup of a Lie group is closed (see Exer. 5), so any subgroup of $\mathrm{SL}(\ell, \mathbb{R})$ that is almost Zariski closed must be closed.

The reader may find it helpful to have some alternative characterizations (see Exer. 6):

(4.2.6) **Remark.**
1) A connected subgroup H of $\mathrm{SL}(\ell, \mathbb{R})$ is almost Zariski closed if and only if it is the identity component of a subgroup that is Zariski closed.
2) A subgroup H of $\mathrm{SL}(\ell, \mathbb{R})$ is almost Zariski closed if and only if it is the union of (finitely many) components of a Zariski closed group.
3) Suppose H has only finitely many connected components. Then H is almost Zariski closed if and only if its identity component H° is almost Zariski closed.
4) Suppose H is a Lie subgroup of $\mathrm{SL}(\ell, \mathbb{R})$. Then H is almost Zariski closed if and only if $\dim \overline{\overline{H}} = \dim H$.

Note that if H is almost Zariski closed, then it is closed, and has only finitely many connected components. Here are two examples to show that the converse is false. (Both examples are closed and connected.) Corollary 4.6.8 below implies that all examples of this phenomenon must be based on similar constructions.

(4.2.7) **Example.**

1) For any irrational number α, let

$$T = \left\{ \begin{bmatrix} t^\alpha & 0 & 0 \\ 0 & t & 0 \\ 0 & 0 & 1/t^{\alpha+1} \end{bmatrix} \middle| \; t \in \mathbb{R}^+ \right\} \subset \mathrm{SL}(3, \mathbb{R}).$$

Then

$$\overline{\overline{T}} = \left\{ \begin{bmatrix} s & 0 & 0 \\ 0 & t & 0 \\ 0 & 0 & 1/(st) \end{bmatrix} \middle| \; s, t \in \mathbb{R} \smallsetminus \{0\} \right\}.$$

Since $\dim T = 1 \neq 2 = \dim \overline{\overline{T}}$, we conclude that T is not almost Zariski closed.

The calculation of $\overline{\overline{T}}$ follows easily from Cor. 4.5.4 below. Intuitively, the idea is simply that, for elements g of T, the relation between $g_{1,1}$ and $g_{2,2}$ is transcendental, not algebraic, so it cannot be captured by a polynomial. Thus, as far as polynomials are concerned, there is no relation at all between $g_{1,1}$ and $g_{2,2}$ — they can vary independently of one another. This independence is reflected in the Zariski closure.

2) Let

$$H = \left\{ \begin{bmatrix} e^t & 0 & 0 & 0 \\ 0 & e^{-t} & 0 & 0 \\ 0 & 0 & 1 & t \\ 0 & 0 & 0 & 1 \end{bmatrix} \middle| \; t \in \mathbb{R} \right\} \subset \mathrm{SL}(4, \mathbb{R}).$$

Then

$$\overline{\overline{H}} = \left\{ \begin{bmatrix} e^s & 0 & 0 & 0 \\ 0 & e^{-s} & 0 & 0 \\ 0 & 0 & 1 & t \\ 0 & 0 & 0 & 1 \end{bmatrix} \middle| \; s, t \in \mathbb{R} \right\} \subset \mathrm{SL}(4, \mathbb{R}).$$

Since $\dim H = 1 \neq 2 = \dim \overline{\overline{H}}$, we conclude that H is not almost Zariski closed.

Formally, the fact that H is not almost-Zariski closed follows from Thm. 4.5.4 below. Intuitively, the transcendental relation between $g_{1,1}$ and $g_{3,4}$ is lost in the Zariski closure.

Exercises for §4.2.

#1. For each subset H of $\mathrm{SL}(\ell, \mathbb{R})$, show there is a unique Zariski closed subset $\overline{\overline{H}}$ of $\mathrm{SL}(\ell, \mathbb{R})$ containing H, such that if C is any Zariski closed subset $\overline{H}$ of $\mathrm{SL}(\ell, \mathbb{R})$ that contains H, then $\overline{\overline{H}} \subset C$.
   [*Hint:* Any intersection of Zariski closed sets is Zariski closed.]

#2. Show that if Z is any subset of an algebraic group G, then the centralizer $C_G(Z)$ is Zariski closed.

#3. Verify 4.2.3(3).
   [*Hint:* Let $\mathcal{Q} = \{x_{1,2}, x_{2,1}\}$. If $Q(x_{1,1}, x_{2,2})$ is a polynomial, such that $Q(t, 1/t) = 0$ for all $t > 0$, then $Q(t, 1/t) = 0$ for all $t \in \mathbb{R}$.]

#4. Show that if H is a **connected** subgroup of $\mathrm{SL}(\ell, \mathbb{R})$, then $\overline{\overline{H}}$ is irreducible.

#5. (*Requires some Lie theory*) Suppose H is a finite-index subgroup of a Lie group G. Show that H is an open subgroup of G. (So H is closed.)
   [*Hint:* There exists $n \in \mathbb{Z}^+$, such that $g^n \in H$ for all $g \in G$. Therefore $\exp(x) = \exp((1/n)x)^n \in H$ for every element x of the Lie algebra of G.]

#6. Verify each part of Rem. 4.2.6.

4.3. Real Jordan decomposition

The real Jordan decomposition writes any matrix as a combination of matrices of three basic types.

(4.3.1) **Definition.** Let $g \in \mathrm{SL}(\ell, \mathbb{R})$.

- g is **unipotent** if 1 is the only eigenvalue of g (over $\mathbb{C}$); in other words, $(g - 1)^\ell = 0$ (see 1.1.7).

- g is **hyperbolic** (or $\mathbb{R}$-**split**) if it is diagonalizable over $\mathbb{R}$, and all of its eigenvalues are positive; that is, if $h^{-1}gh$ is a diagonal matrix with no negative entries, for some $h \in \mathrm{SL}(\ell, \mathbb{R})$.

- g is **elliptic** if it is diagonalizable over $\mathbb{C}$, and all of its eigenvalues are of absolute value 1.

(4.3.2) **Example.** For all $t \in \mathbb{R}$:

1) $\begin{bmatrix} 1 & 0 \\ t & 1 \end{bmatrix}$ is unipotent,

2) $\begin{bmatrix} e^t & 0 \\ 0 & e^{-t} \end{bmatrix}$ is hyperbolic,

3) $\begin{bmatrix} \cos t & \sin t \\ -\sin t & \cos t \end{bmatrix}$ is elliptic (see Exer. 2).

See Exer. 3 for an easy way to tell whether an element of $SL(2, \mathbb{R})$ is unipotent, hyperbolic, or elliptic.

(4.3.3) **Proposition** (Real Jordan decomposition). *For any $g \in SL(\ell, \mathbb{R})$, there exist unique $g_u, g_h, g_e \in SL(\ell, \mathbb{R})$, such that*

1) $g = g_u g_h g_e$,
2) g_u *is unipotent,*
3) g_h *is hyperbolic,*
4) g_e *is elliptic, and*
5) $g_u, g_h,$ *and g_e all commute with each other.*

Proof. (Existence) The usual Jordan decomposition of Linear Algebra (also known as "Jordan Canonical Form") implies there exist $h \in SL(\ell, \mathbb{C})$, a nilpotent matrix N, and a diagonal matrix D, such that $h^{-1}gh = N + D$, and N commutes with D. This is an additive decomposition. By factoring out D, we obtain a multiplicative decomposition:

$$h^{-1}gh = (ND^{-1} + I)D = uD,$$

where $u = ND^{-1} + I$ is unipotent (because $u - I = ND^{-1}$ is nilpotent, since N commutes with D^{-1}).

Now, because any complex number z has a (unique) polar form $z = re^{i\theta}$, we may write $D = D_h D_e$, where D_h is hyperbolic, D_e is elliptic, and both matrices are diagonal, so they commute with each other (and, from the structure of the Jordan Canonical Form, they both commute with N). Conjugating by h^{-1}, we obtain

$$g = h(uD_h D_e)h^{-1} = g_u g_h g_e,$$

where $g_u = huh^{-1}$, $g_h = hD_h h^{-1}$, and $g_e = hD_e h^{-1}$. This is the desired decomposition.

(Uniqueness) The uniqueness of the decomposition is, perhaps, not so interesting to the reader, so we relegate it to the exercises (see Exers. 5 and 6). Uniqueness is, however, often of vital importance. For example, it can be used to address a technical difficulty that was ignored in the above proof: from our construction, it appears that the matrices g_u, g_h, and g_e may have complex entries, not real. However, using an overline to denote complex conjugation, we have $\overline{g} = \overline{g_u}\,\overline{g_h}\,\overline{g_e}$. Since $\overline{g} = g$, the uniqueness of the decomposition implies $\overline{g_u} = g_u$, $\overline{g_h} = g_h$, and $\overline{g_e} = g_e$. Therefore, $g_u, g_h, g_e \in \mathrm{SL}(\ell, \mathbb{R})$, as desired. $\square$

The uniqueness of the Jordan decomposition implies, for $g, h \in \mathrm{SL}(\ell, \mathbb{R})$, that if g commutes with h, then the Jordan components g_u, g_h, and g_e commute with h (see also Exer. 5). In other words, if the centralizer $C_{\mathrm{SL}(\ell,\mathbb{R})}(h)$ contains g, then it must also contain the Jordan components of g. Because the centralizer is Zariski closed (see Exer. 4.2#2), this is a special case of the following important result.

(4.3.4) Theorem. *If*

- *G is a Zariski closed subgroup of* $\mathrm{SL}(\ell, \mathbb{R})$, *and*
- *$g \in G$,*

then $g_u, g_h, g_e \in G$.

We postpone the proof to §4.5.

As mentioned at the start of the chapter, we should assume that homomorphisms are polynomial functions. (But some other types of functions will be allowed to be more general rational functions, which are not defined when the denominator is 0.)

(4.3.5) Definition. Let H be a subset of $\mathrm{SL}(\ell, \mathbb{R})$.

1) A function $\phi \colon H \to \mathbb{R}$ is a *polynomial* (or is *regular*) if there exists $Q \in \mathbb{R}[x_{1,1}, \ldots, x_{\ell,\ell}]$, such that $\phi(h) = Q(h)$ for all $h \in H$.

2) A real-valued function ψ defined a subset of H is *rational* if there exist polynomials $\phi_1, \phi_2 \colon H \to \mathbb{R}$, such that

   (a) the domain of ψ is $\{\, h \in H \mid \phi_2(h) \neq 0 \,\}$, and

   (b) $\psi(h) = \phi_1(h)/\phi_2(h)$ for all h in the domain of ψ.

3) A function $\phi \colon H \to \mathrm{SL}(n, \mathbb{R})$ is a *polynomial* if, for each $1 \le i, j \le n$, the matrix entry $\phi(h)_{i,j}$ is a polynomial function of $h \in H$. Similarly, ψ is *rational* if each $\psi(h)_{i,j}$ is a rational function of $h \in H$.

We now show that any polynomial homomorphism respects the real Jordan decomposition; that is, $\rho(g_u) = \rho(g)_u$, $\rho(g_h) = \rho(g)_h$, and $\rho(g_e) = \rho(g)_e$.

(4.3.6) Corollary. *Suppose*

- *G is a real algebraic group, and*
- *$\rho \colon G \to \mathrm{SL}(m, \mathbb{R})$ is a polynomial homomorphism.*

Then:

1) *If u is any unipotent element of G, then $\rho(u)$ is a unipotent element of* SL$(m, \mathbb{R})$.

2) *If a is any hyperbolic element of G, then $\rho(a)$ is a hyperbolic element of* SL$(m, \mathbb{R})$.

3) *If k is any elliptic element of G, then $\rho(k)$ is an elliptic element of* SL$(m, \mathbb{R})$.

Proof. Note that the graph of ρ is a Zariski closed subgroup of $G \times H$ (see Exer. 15).

We prove only (1); the others are similar. Since u is unipotent, we have $u_u = u$, $u_h = e$, and $u_e = e$. Therefore, the real Jordan decomposition of $(u, \rho(u))$ is

$$(u, \rho(u)) = (u, \rho(u)_u)(e, \rho(u)_h)(e, \rho(u)_e).$$

Since $(u, \rho(u)) \in \operatorname{graph} \rho$, Thm. 4.3.4 implies

$$(u, \rho(u)_u) = (u, \rho(u))_u \in \operatorname{graph} \rho.$$

Let $y = \rho(u)_u$. Since $(u, y) \in \operatorname{graph} \rho$, we have $\rho(u) = y$. Hence $\rho(u) = \rho(u)_u$ is unipotent. $\qquad \square$

Exercises for §4.3.

#1. Show that every element of $\mathbb{U}_\ell$ is unipotent.

#2. Show $\begin{bmatrix} \cos t & \sin t \\ -\sin t & \cos t \end{bmatrix}$ is an elliptic element of SL$(2, \mathbb{R})$, for every $t \in \mathbb{R}$.

#3. Let $g \in$ SL$(2, \mathbb{R})$. Recall that trace g is the sum of the diagonal entries of g. Show:

(a) g is unipotent if and only if trace $g = 2$.

(b) g is hyperbolic if and only if trace $g > 2$.

(c) g is elliptic if and only if $-2 <$ trace $g < 2$.

(d) g is neither unipotent, hyperbolic, nor elliptic if and only if trace $g \leq -2$.

#4. Suppose g and h are elements of SL$(\ell, \mathbb{R})$, such that $gh = hg$. Show:

(a) If g and h are unipotent, then gh is unipotent.

(b) If g and h are hyperbolic, then gh is hyperbolic.

(c) If g and h are elliptic, then gh is elliptic.

#5. Suppose $g, g_u, g_h, g_e \in$ SL$(\ell, \mathbb{C})$, and these matrices are as described in the conclusion of Prop. 4.3.3. Show (without using the Jordan decomposition or any of its properties) that if $x \in$ SL$(\ell, \mathbb{C})$, and x commutes with g, then x also commutes with each of g_u, g_h, and g_e.

[*Hint:* Passing to a conjugate, assume g_h and g_e are diagonal. We have $g_h^{-n} x g_h^n = (g_u g_e)^n x (g_u g_e)^{-n}$. Since each matrix entry of the LHS is an exponential function of n, but each matrix entry on the RHS grows at most polynomially, we see that the LHS must be constant. So x commutes with g_h. Then $g_u^{-n} x g_u^n = g_e^n x g_e^{-n}$. Since a bounded polynomial must be constant, we see that x commutes with g_u and g_e.]

#6. Show that the real Jordan decomposition is unique.
[*Hint:* If $g = g_u g_h g_e = g_u' g_h' g_e'$, then $g_u^{-1} g_u' = g_h g_e (g_h' g_e')^{-1}$ is both unipotent and diagonalizable over $\mathbb{C}$ (this requires Exer. 5). Therefore $g_u = g_u'$. Similarly, $g_h = g_h'$ and $g_e = g_e'$.]

#7. Suppose $g \in \mathrm{SL}(\ell, \mathbb{R})$, $v \in \mathbb{R}^\ell$, and v is an eigenvector for g. Show that v is also an eigenvector for g_u, g_h, and g_e. [*Hint:* Let W be the eigenspace corresponding to the eigenvalue λ associated to v. Because g_u, g_h, and g_e commute with g, they preserve W. The Jordan decomposition of $g|_W$, the restriction of g to W, is $(g|_W)_u (g|_W)_h (g|_W)_e$.]

#8. Show that any commuting set of diagonalizable matrices can be diagonalized simultaneously. More precisely, suppose
   - $S \subset \mathrm{SL}(\ell, \mathbb{R})$,
   - each $s \in S$ is hyperbolic, and
   - the elements of S all commute with each other.

Show there exists $h \in \mathrm{SL}(\ell, \mathbb{R})$, such that every element of $h^{-1} S h$ is diagonal.

#9. Suppose G is an subgroup of $\mathrm{SL}(\ell, \mathbb{R})$ that is almost Zariski closed.
   (a) For $i(g) = g^{-1}$, show that i is a polynomial function from G to G.
   (b) For $m(g, h) = gh$, show that m is a polynomial function from $G \times G$ to G. (Note that $G \times G$ can naturally be realized as a subgroup of $\mathrm{SL}(2\ell, \mathbb{R})$ that is almost Zariski closed.)
   [*Hint:* Cramer's Rule provides a polynomial formula for the inverse of a matrix of determinant one. The usual formula for the product of two matrices is a polynomial.]

#10. Show that if
   - $f \colon \mathrm{SL}(\ell, \mathbb{R}) \to \mathrm{SL}(m, \mathbb{R})$ is a polynomial, and
   - H is a Zariski closed subgroup of $\mathrm{SL}(m, \mathbb{R})$,

then $f^{-1}(H)$ is Zariski closed.

#11. Show that if H is any subgroup of $\mathrm{SL}(\ell, \mathbb{R})$, then $\overline{\overline{H}}$ is also a subgroup of $\mathrm{SL}(\ell, \mathbb{R})$.
[*Hint:* Exercises 9 and 10.]

#12. Show that if H is a connected Lie subgroup of $\mathrm{SL}(\ell, \mathbb{R})$, then the normalizer $N_{\mathrm{SL}(\ell,\mathbb{R})}(H)$ is Zariski closed.
[*Hint:* The homomorphism Ad: $\mathrm{SL}(\ell, \mathbb{R}) \to \mathrm{SL}(\mathfrak{sl}(\ell, \mathbb{R}))$ is a polynomial.]

#13. Show that if G is any connected subgroup of $\mathrm{SL}(\ell, \mathbb{R})$, then G is a *normal* subgroup of $\overline{\overline{G}}$.

#14. There is a natural embedding of $\mathrm{SL}(\ell, \mathbb{R}) \times \mathrm{SL}(m, \mathbb{R})$ in $\mathrm{SL}(\ell + m, \mathbb{R})$. Show that if G and H are Zariski closed subgroups of $\mathrm{SL}(\ell, \mathbb{R})$ and $\mathrm{SL}(m, \mathbb{R})$, respectively, then $G \times H$ is Zariski closed in $\mathrm{SL}(\ell + m, \mathbb{R})$.

#15. Suppose G is a Zariski closed subgroup of $\mathrm{SL}(\ell, \mathbb{R})$, and $\rho: G \to \mathrm{SL}(m, \mathbb{R})$ is a polynomial homomorphism. There is a natural embedding of the graph of ρ in $\mathrm{SL}(\ell + m, \mathbb{R})$ (cf. Exer. 14). Show that the graph of ρ is Zariski closed.

4.4. Structure of almost-Zariski closed groups

The main result of this section is that any algebraic group can be decomposed into subgroups of three basic types: unipotent, torus, and semisimple (see Thm. 4.4.7).

(4.4.1) **Definition.**

- A subgroup U of $\mathrm{SL}(\ell, \mathbb{R})$ is **unipotent** if and only if it is conjugate to a subgroup of $\mathbb{U}_\ell$.

- A subgroup T of $\mathrm{SL}(\ell, \mathbb{R})$ is a **torus** if
    - T is conjugate (over $\mathbb{C}$) to a group of diagonal matrices; that is, $h^{-1}Th$ consists entirely of diagonal matrices, for some $h \in \mathrm{SL}(\ell, \mathbb{C})$),
    - T is connected, and
    - T is almost Zariski closed.

  (We have required tori to be connected, but this requirement should be relaxed slightly; any subgroup of $\overline{\overline{T}}$ that contains T may also be called a torus.)

- A closed subgroup L of $\mathrm{SL}(\ell, \mathbb{R})$ is **semisimple** if its identity component $L°$ has no nontrivial, connected, abelian, normal subgroups.

(4.4.2) **Remark.** Here are alternative characterizations of unipotent groups and tori:

1) (Engel's Theorem) A subgroup U of $\mathrm{SL}(\ell, \mathbb{R})$ is **unipotent** if and only if every element of U is unipotent (see Exer. 5).
2) A connected subgroup T of $\mathrm{SL}(\ell, \mathbb{R})$ is a torus if and only if

   - T is abelian,
   - each individual element of T is diagonalizable (over $\mathbb{C}$), and
   - T is almost Zariski closed

   (see Exer. 4.3#8).

Unipotent groups and tori are fairly elementary, but the semisimple groups are more difficult to understand. The following fundamental theorem of Lie theory reduces their study to simple groups (which justifies their name).

(4.4.3) Definition. A group G is **almost simple** if it has no infinite, proper, normal subgroups.

(4.4.4) Theorem. *Let L be a connected, semisimple subgroup of $\mathrm{SL}(\ell, \mathbb{R})$. Then, for some n, there are closed, connected subgroups $S_1, \ldots, S_n$ of L, such that*

1) *each S_i is almost simple, and*
2) *L is isomorphic to (a finite cover of) $S_1 \times \cdots \times S_n$.*

The almost-simple groups have been classified by using the theory known as "roots and weights." We merely provide some typical examples, without proof.

(4.4.5) Example.

1) $\mathrm{SL}(\ell, \mathbb{R})$ is almost simple (if $\ell \geq 2$).
2) If Q is a quadratic form on $\mathbb{R}^\ell$ that is nondegenerate (see Defn. 1.2.1), and $\ell \geq 3$, then $\mathrm{SO}(Q)$ is semisimple (and it is almost simple if, in addition, $n \neq 4$). (For $\ell = 2$, the groups $\mathrm{SO}(2)$ and $\mathrm{SO}(1, 1)$ are tori, not semisimple (see Exer. 1).)

From the above almost-simple groups, it is easy to construct numerous semisimple groups. One example is

$$\mathrm{SL}(3, \mathbb{R}) \times \mathrm{SL}(7, \mathbb{R}) \times \mathrm{SO}(6) \times \mathrm{SO}(4, 7).$$

The following structure theorem is one of the major results in the theory of algebraic groups.

(4.4.6) Definition. Recall that a Lie group G is a **semidirect product** of closed subgroups A and B (denoted $G = A \ltimes B$) if

1) $G = AB$,
2) B is a normal subgroup of G, and

3) $A \cap B = \{e\}$.

(In this case, the map $(a, b) \mapsto ab$ is a diffeomorphism from $A \times B$ onto G. However, it is not a group isomorphism (or even a homomorphism) unless every element of A commutes with every element of B.)

(4.4.7) Theorem. *Let G be a connected subgroup of* $\mathrm{SL}(\ell, \mathbb{R})$ *that is almost Zariski closed. Then there exist:*

- *a semisimple subgroup L of G,*
- *a torus T in G, and*
- *a unipotent subgroup U of G,*

such that

1) *$G = (LT) \ltimes U$,*

2) *L, T, and U are almost Zariski closed, and*

3) *L and T centralize each other, and have finite intersection.*

Sketch of proof (*requires some Lie theory*). Let R be the radical of G, and let L be a Levi subgroup of G; thus, R is solvable, L is semisimple, $LR = G$, and $L \cap R$ is discrete (see 4.9.15). From the Lie-Kolchin Theorem (4.9.17), we know that R is conjugate (over $\mathbb{C}$) to a group of lower-triangular matrices. By working in $\mathrm{SL}(\ell, \mathbb{C})$, let us assume, for simplicity, that R itself is lower triangular. That is, $R \subset \mathbb{D}_\ell \ltimes \mathbb{U}_\ell$.

Let $\pi \colon \mathbb{D}_\ell \ltimes \mathbb{U}_\ell \to \mathbb{D}_\ell$ be the natural projection. It is not difficult to see that there exists $r \in R$, such that $\pi(R) \subset \overline{\langle \pi(r) \rangle}$ (by using (4.4.12) and (4.5.4)). Let

$$T = \overline{\langle r_s r_e \rangle}^{\circ} \text{ and } U = R \cap \mathbb{U}_\ell.$$

Because $\pi(r_s r_e) = \pi(r)$, we have $\pi(R) \subset \pi(T)$, so, for any $g \in R$, there exists $t \in T$, such that $\pi(t) = \pi(g)$. Then $\pi(t^{-1}g) = e$, so $t^{-1}g \in U$. Therefore $g \in tU \subset T \ltimes U$. Since $g \in R$ is arbitrary, we conclude that

$$R = T \ltimes U.$$

This yields the desired decomposition $G = (LT) \ltimes U$.   □

(4.4.8) Remark. The subgroup U of (4.4.7) is the **unique** maximal unipotent normal subgroup of G. It is called the **unipotent radical** of G.

It is obvious (from the Jordan decomposition) that every element of a compact real algebraic group is elliptic. We conclude this section by recording (without proof) the fact that this characterizes the compact real algebraic groups.

(4.4.9) Theorem. *An almost-Zariski closed subgroup of* $\mathrm{SL}(\ell, \mathbb{R})$ *is compact if and only if all of its elements are elliptic.*

(4.4.10) Corollary.

1) *A nontrivial unipotent subgroup U of* $\mathrm{SL}(\ell, \mathbb{R})$ *is never compact.*

2) *A torus T in* $\mathrm{SL}(\ell, \mathbb{R})$ *is compact if and only if none of its nontrivial elements are hyperbolic.*

3) *A connected, semisimple subgroup L of* $\mathrm{SL}(\ell, \mathbb{R})$ *is compact if and only if it has no nontrivial unipotent elements (also, if and only if it has no nontrivial hyperbolic elements).*

We conclude this section with two basic results about tori.

(4.4.11) Definition. A torus T is **hyperbolic** (or $\mathbb{R}$-**split**) if every element of T is hyperbolic.

(4.4.12) Corollary. *Any connected torus T has a unique decomposition into a direct product $T = T_h \times T_c$, where*

1) *T_h is a hyperbolic torus, and*

2) *T_c is a compact torus.*

Proof. Let
$$T_h = \{\, g \in T \mid g \text{ is hyperbolic} \,\}$$
and
$$T_c = \{\, g \in T \mid g \text{ is elliptic} \,\}.$$
Because T is abelian, it is easy to see that T_h and T_c are subgroups of T (see Exer. 4.3#4). It is immediate from the real Jordan decomposition that $T = T_h \times T_c$.

All that remains is to show that T_h and T_c are almost Zariski closed.

(T_h) Since T_h is a set of commuting matrices that are diagonalizable over $\mathbb{R}$, there exists $h \in \mathrm{SL}(\ell, \mathbb{R})$, such that $h^{-1} T_h h \subset \mathbb{D}_\ell$ (see Exer. 4.3#8). Hence, $T_h = T \cap (h\mathbb{D}_\ell h^{-1})$ is almost Zariski closed.

(T_e) Let

$\mathbb{D}_\ell^{\mathbb{C}}$ be the group of diagonal matrices in $\mathrm{SL}(\ell, \mathbb{C})$,

and
$$C = \left\{\, g \in \mathbb{D}_\ell^{\mathbb{C}} \;\middle|\; \begin{array}{l} \text{every eigenvalue of } g \\ \text{has absolute value } 1 \end{array} \right\}.$$

Because T is a torus, there exists $h \in \mathrm{SL}(\ell, \mathbb{C})$, such that $h^{-1} T h \subset \mathbb{D}_\ell^{\mathbb{C}}$. Then $T_c = T \cap hCh^{-1}$ is compact. So it is Zariski closed (see Prop. 4.6.1 below). $\square$

A (real) ***representation*** of a group is a homomorphism into $SL(m, \mathbb{R})$, for some m. The following result provides an explicit description of the representations of any hyperbolic torus.

(4.4.13) Corollary. *Suppose*

- *T is a (hyperbolic) torus that consists of diagonal matrices in $SL(\ell, \mathbb{R})$, and*

- *$\rho: T \to SL(m, \mathbb{R})$ is any polynomial homomorphism.*

Then there exists $h \in SL(n, \mathbb{R})$, such that, letting

$$\rho'(t) = h^{-1} \rho(t) h \quad for \ t \in T,$$

we have:

1) *$\rho'(T) \subset \mathbb{D}_m$, and*

2) *For each j with $1 \le j \le m$, there are integers $n_1, \ldots, n_\ell$, such that*

$$\rho'(t)_{j,j} = t_{1,1}^{n_1} t_{2,2}^{n_2} \cdots t_{\ell,\ell}^{n_\ell}$$

for all $t \in T$.

Proof. (1) Since $\rho(T)$ is a set of commuting matrices that are diagonalizable over $\mathbb{R}$, there exists $h \in SL(m, \mathbb{R})$, such that $h^{-1}\rho(T)h \subset \mathbb{D}_m$ (see Exer. 4.3#8).

(2) For each j, $\rho'(t)_{j,j}$ defines a polynomial homomorphism from T to $\mathbb{R}^+$. With the help of Lie theory, it is not difficult to see that any such homomorphism is of the given form (see Exer. 4.9#6). $\square$

Exercises for §4.4.

#1. Show:

(a) $SO(2)$ is a compact torus, and

(b) $SO(1, 1)°$ is a hyperbolic torus.
[*Hint:* We have

$$SO(2) = \left\{ \begin{bmatrix} \cos\theta & \sin\theta \\ -\sin\theta & \cos\theta \end{bmatrix} \right\} \text{ and } SO(1,1) = \left\{ \pm \begin{bmatrix} \cosh t & \sinh t \\ \sinh t & \cosh t \end{bmatrix} \right\},$$

where $\cosh t = (e^t + e^{-t})/2$ and $\sinh t = (e^t - e^{-t})/2$.]

#2. Show:

(a) The set of unipotent elements of $SL(\ell, \mathbb{R})$ is Zariski closed.

(b) If U is a unipotent subgroup of $SL(\ell, \mathbb{R})$, then $\overline{\overline{U}}$ is also unipotent.

#3. Prove the easy direction ($\Rightarrow$) of Thm. 4.4.9.

#4. Assume that Thm. 4.4.9 has been proved for semisimple groups. Prove the general case.
[*Hint:* Use Thm. 4.4.7.]

#5. (*Advanced*) Prove Engel's Theorem 4.4.2(1).
[*Hint:* ($\Leftarrow$) It suffices to show that U fixes some nonzero vector v. (For then we may consider the action of U on $\mathbb{R}^{\ell}/\mathbb{R}v$, and complete the proof by induction on ℓ.) There is no harm in working over $\mathbb{C}$, rather than $\mathbb{R}$, and we may assume there are no U-invariant subspaces of $\mathbb{C}^{\ell}$. Then a theorem of Burnside states that every $\ell \times \ell$ matrix M is a linear combination of elements of U. Hence, for any $u \in U$, trace(uM) = trace M. Since M is arbitrary, we conclude that $u = I$.]

4.5. Chevalley's Theorem and applications

(4.5.1) **Notation.** For a map $\rho \colon G \to Z$ and $g \in G$, we often write g^{ρ} for the image of g under ρ. That is, g^{ρ} is another notation for $\rho(g)$.

(4.5.2) **Proposition** (Chevalley's Theorem). *A subgroup H of a real algebraic group G is Zariski closed if and only if, for some m, there exist*

- *a polynomial homomorphism $\rho \colon G \to \mathrm{SL}(m, \mathbb{R})$, and*
- *a vector $v \in \mathbb{R}^m$,*

such that $H = \{\, h \in G \mid v h^{\rho} \in \mathbb{R}v \,\}$.

Proof. ($\Leftarrow$) This follows easily from Eg. 4.1.2(7) and Exer. 4.3#10.

($\Rightarrow$) There is no harm in assuming $G = \mathrm{SL}(\ell, \mathbb{R})$. There is a **finite** subset Q of $\mathbb{R}[x_{1,1}, \ldots, x_{\ell,\ell}]$, such that $H = \mathrm{Var}(Q)$ (see Exer. 4.1#7c). Choose $d \in \mathbb{Z}^+$, such that $\deg Q < d$ for all $Q \in Q$, and let

- $V = \{\, Q \in \mathbb{R}[x_{1,1}, \ldots, x_{\ell,\ell}] \mid \deg Q < d \,\}$ and
- $W = \{\, Q \in V \mid Q(h_{i,j}) = 0 \text{ for all } h \in H \,\}$.

Thus, we have $H = \bigcap_{Q \in W} \mathrm{Var}(\{Q\})$.

There is a natural homomorphism ρ from $\mathrm{SL}(\ell, \mathbb{R})$ to the group $\mathrm{SL}(V)$ of (special) linear transformations on V, defined by

$$(Q g^{\rho})(x_{i,j}) = Q((gx)_{i,j}) \tag{4.5.3}$$

(see Exer. 2a). Note that we have $\mathrm{Stab}_{\mathrm{SL}(\ell,\mathbb{R})}(W) = H$ (see Exer. 2b). By taking a basis for V, we may think of ρ as a polynomial homomorphism into $\mathrm{SL}(\dim V, \mathbb{R})$ (see Exer. 2c). Then this is almost exactly what we want; the only problem is that, instead of a 1-dimensional space $\mathbb{R}v$, we have the space W of (possibly) larger dimension.

To complete the proof, we convert W into a 1-dimensional space, by using a standard trick of multilinear algebra. For $k = \dim W$, we let

$$V' = \textstyle\bigwedge^k V \text{ and } W' = \textstyle\bigwedge^k W \subset V',$$

where $\bigwedge^k V$ denotes the kth exterior power of V. Now ρ naturally induces a polynomial homomorphism $\rho' \colon \mathrm{SL}(\ell, \mathbb{R}) \to \mathrm{SL}(V')$, and, for this action, $H = \mathrm{Stab}_{\mathrm{SL}(\ell,\mathbb{R})}(W)$ (see Exer. 3). By choosing a basis for V', we can think of ρ' as a homomorphism into $\mathrm{SL}\left(\binom{\dim V}{k}, \mathbb{R}\right)$. Since $\dim W' = \binom{\dim W}{k} = 1$, we obtain the desired conclusion (with ρ' in the place of ρ) by letting v be any nonzero vector in W'. $\qquad\square$

Proof of Thm. 4.3.4. From Chevalley's Theorem (4.5.2), we know there exist

- a polynomial homomorphism $\rho \colon \mathrm{SL}(\ell, \mathbb{R}) \to \mathrm{SL}(m, \mathbb{R})$, for some m, and
- a vector $v \in \mathbb{R}^m$,

such that $G = \{ g \in \mathrm{SL}(\ell, \mathbb{R}) \mid v g^\rho \in \mathbb{R}v \}$. Furthermore, from the explicit description of ρ in the proof of Prop. 4.5.2, we see that it satisfies the conclusions of Cor. 4.3.6 with $\mathrm{SL}(\ell, \mathbb{R})$ in the place of G (cf. Exer. 4). Thus, for any $g \in \mathrm{SL}(\ell, \mathbb{R})$, we have

$$(g_u)^\rho = (g^\rho)_u, \ (g_h)^\rho = (g^\rho)_h, \text{ and } (g_e)^\rho = (g^\rho)_e.$$

For any $g \in G$, we have $v g^\rho \in \mathbb{R}v$. In other words, v is an eigenvector for g^ρ. Then v is also an eigenvector for $(g^\rho)_u$ (see Exer. 4.3#7). Since $(g_u)^\rho = (g^\rho)_u$, this implies $v(g_u)^\rho \in \mathbb{R}v$, so $g_u \in G$. By the same argument, $g_h \in G$ and $g_e \in G$. $\qquad\square$

Chevalley's Theorem yields an explicit description of the hyperbolic tori.

(4.5.4) Corollary. *Suppose T is a connected group of diagonal matrices in $\mathrm{SL}(\ell, \mathbb{R})$, and let $d = \dim T$. Then T is almost Zariski closed if and only if there are linear functionals $\lambda_1, \ldots, \lambda_\ell \colon \mathbb{R}^d \to \mathbb{R}$, such that*

$$1) \ T = \left\{ \begin{bmatrix} e^{\lambda_1(x)} & & & & \\ & e^{\lambda_2(x)} & & & \\ & & \ddots & & \\ & & & e^{\lambda_\ell(x)} \end{bmatrix} \ \middle| \ x \in \mathbb{R}^d \right\}, \text{ and}$$

2) *for each i, there are **integers** $n_1, \ldots, n_d$, such that*

$$\lambda_i(x_1, \ldots, x_d) = n_1 x_1 + \cdots + n_d x_d \text{ for all } x \in \mathbb{R}^d.$$

Proof. Combine Prop. 4.5.2 with Cor. 4.4.13 (see Exer. 5). $\qquad\square$

Exercises for §4.5.

#1. Suppose $Q \in \mathbb{R}[x_{1,1}, \ldots, x_{\ell,\ell}]$ and $g \in \mathrm{SL}(\ell, \mathbb{R})$.

   • Let $\phi \colon \mathrm{SL}(\ell, \mathbb{R}) \to \mathbb{R}$ be the polynomial function corresponding to Q, and

   • define $\phi' \colon \mathrm{SL}(\ell, \mathbb{R}) \to \mathbb{R}$ by $\phi'(x) = \phi(gx)$.

   Show there exists $Q' \in \mathbb{R}[x_{1,1}, \ldots, x_{\ell,\ell}]$, with $\deg Q' = \deg Q$, such that ϕ' is the polynomial function corresponding to Q'.
   [*Hint:* For fixed g, the matrix entries of gh are linear functions of h.]

#2. Define $\rho \colon \mathrm{SL}(\ell, \mathbb{R}) \to \mathrm{SL}(V)$ as in Eq. (4.5.3).

   (a) Show ρ is a group homomorphism.

   (b) For the subspace W defined in the proof of Prop. 4.5.2, show $H = \mathrm{Stab}_{\mathrm{SL}(\ell, \mathbb{R})}(W)$.

   (c) By taking a basis for V, we may think of ρ as a map into $\mathrm{SL}(\dim V, \mathbb{R})$. Show ρ is a polynomial.
   [*Hint:* (2b) We have $Q \subset W$.]

#3. Suppose

   • W is a subspace of a real vector space V,

   • g is an invertible linear transformation on V, and

   • $k = \dim W$.

   Show $\bigwedge^k (Wg) = \bigwedge^k W$ if and only if $Wg = W$.

#4. Define $\rho \colon \mathrm{SL}(\ell, \mathbb{R}) \to \mathrm{SL}(V)$ as in Eq. (4.5.3).

   (a) Show that if g is hyperbolic, then $\rho(g)$ is hyperbolic.

   (b) Show that if g is elliptic, then $\rho(g)$ is elliptic.

   (c) Show that if g is unipotent, then $\rho(g)$ is unipotent.
   [*Hint:* (4a,4b) If g is diagonal, then any monomial is an eigenvector of g^ρ.]

#5. Prove Cor. 4.5.4.

4.6. Subgroups that are almost Zariski closed

We begin the section with some results that guarantee certain types of groups are almost Zariski closed.

(4.6.1) Proposition. *Any compact subgroup of* $\mathrm{SL}(\ell, \mathbb{R})$ *is Zariski closed.*

Proof. Suppose C is a compact subgroup of $SL(\ell, \mathbb{R})$, and g is an element of $SL(\ell, \mathbb{R}) \smallsetminus C$. It suffices to find a polynomial ϕ on $SL(\ell, \mathbb{R})$, such that $\phi(C) = 0$, but $\phi(g) \neq 0$.

The sets C and Cg are compact and disjoint, so, for any $\epsilon > 0$, the Stone-Weierstrass Theorem implies there is a polynomial ϕ_0, such that $\phi_0(c) < \epsilon$ and $\phi_0(cg) > 1 - \epsilon$ for all $c \in C$. (For our purposes, we may choose any $\epsilon < 1/2$.) For each $c \in C$, let $\phi_c(x) = \phi(cx)$, so ϕ_c is a polynomial of the same degree as ϕ_0 (see Exer. 4.5#1). Define $\overline{\phi} \colon SL(\ell, \mathbb{R}) \to \mathbb{R}$ by averaging over $c \in C$:

$$\overline{\phi}(x) = \int_C \phi_c(x)\, dc,$$

where dc is the Haar measure on C, normalized to be a probability measure. Then

1) $\overline{\phi}(c) < \epsilon$ for $c \in C$,

2) $\overline{\phi}(g) > 1 - \epsilon$,

3) $\overline{\phi}$ is constant on C (because Haar measure is invariant), and

4) $\overline{\phi}$ is a polynomial function (each of its coefficients is the average of the corresponding coefficients of the ϕ_c's).

Now let $\phi(x) = \overline{\phi}(x) - \overline{\phi}(c)$ for any $c \in C$. $\qquad\square$

(4.6.2) **Proposition.** *If U is a connected, unipotent subgroup of $SL(\ell, \mathbb{R})$, then U is Zariski closed.*

Proof (*requires some Lie theory*). By passing to a conjugate, we may assume $U \subset \mathbb{U}_\ell$. The Lie algebra $\mathfrak{U}_\ell$ of $\mathbb{U}_\ell$ is the space of strictly lower-triangular matrices (see Exer. 1). Because $A^\ell = 0$ for $A \in \mathfrak{U}_\ell$, the exponential map

$$\exp(A) = I + A + \frac{1}{2}A^2 + \cdots + \frac{1}{(\ell - 1)!}A^{\ell-1}$$

is a polynomial function on $\mathfrak{U}_\ell$, and its inverse, the logarithm map

$$\log(I + N) = N - \frac{1}{2}N^2 + \frac{1}{3}N^3 \pm \cdots \pm \frac{1}{\ell - 1}N^{\ell-1},$$

is a polynomial function on $\mathbb{U}_\ell$.

Therefore $\exp$ is a bijection from $\mathfrak{U}_\ell$ onto $\mathbb{U}_\ell$, so $U = \exp\mathfrak{u}$, where $\mathfrak{u}$ is the Lie algebra of U. This means

$$U = \{\, u \in \mathbb{U}_\ell \mid \log u \in \mathfrak{u} \,\}.$$

Since log is a polynomial function (and $\mathfrak{u}$, being a linear subspace, is defined by polynomial equations — in fact, linear equations), this implies that U is defined by polynomial equations. Therefore, U is Zariski closed. $\square$

The following result is somewhat more difficult; we omit the proof.

(4.6.3) Theorem. *If L is any connected, semisimple subgroup of $SL(\ell, \mathbb{R})$, then L is almost Zariski closed.*

The following three results show that being almost Zariski closed is preserved by certain natural operations. We state the first without proof.

(4.6.4) Proposition. *If A and B are almost-Zariski closed subgroups of $SL(\ell, \mathbb{R})$, such that AB is a subgroup, then AB is almost Zariski closed.*

(4.6.5) Corollary. *If G and H are almost Zariski closed, and ρ is a polynomial homomorphism from G to H, then the image $\rho(G)$ is an almost-Zariski closed subgroup of H.*

Proof. By passing to a finite-index subgroup, we may assume G is connected. Write $G = (TL) \ltimes U$, as in Thm. 4.4.7. From Prop. 4.6.4, it suffices to show that $\rho(U)$, $\rho(L)$, and $\rho(T)$ are almost Zariski closed. The subgroups $\rho(U)$ and $\rho(L)$ are handled by Prop. 4.6.2 and Thm. 4.6.3.

Write $T = T_h \times T_c$, where T_h is hyperbolic and T_c is compact (see Cor. 4.4.12). Then $\rho(T_c)$, being compact, is Zariski closed (see Prop. 4.6.1). The subgroup $\rho(T_h)$ is handled easily by combining Cors. 4.5.4 and 4.4.13 (see Exer. 2). $\square$

(4.6.6) Corollary. *If G is any connected subgroup of $SL(\ell, \mathbb{R})$, then the commutator subgroup $[G, G]$ is almost Zariski closed.*

Proof. Write $G = (LT) \ltimes U$, as in Thm. 4.4.7. Because T is abelian and $[L, L] = L$, we see that $[G, G]$ is a (connected subgroup of $L \ltimes U$ that contains L. Hence $[G, G] = L \ltimes \check{U}$, where $\check{U} = [G, G] \cap U$ (see Exer. 3). Furthermore, since $[G, G]$ is connected, we know $\check{U}$ is connected, so $\check{U}$ is Zariski closed (see Prop. 4.6.2). Since L is almost Zariski closed (see Thm. 4.6.3), this implies $[G, G] = L\check{U}$ is almost Zariski closed (see Prop. 4.6.4), as desired. $\square$

(4.6.7) Corollary. *If G is any connected subgroup of $SL(\ell, \mathbb{R})$, then $[\overline{\overline{G}}, \overline{\overline{G}}] = [G, G]$, so $\overline{\overline{G}}/G$ is abelian.*

Proof. Define $c\colon \overline{\overline{G}} \times \overline{\overline{G}} \to \overline{\overline{G}}$ by $c(g,h) = g^{-1}h^{-1}gh = [g,h]$. Then c is a polynomial (see Exer. 4.3#9). Since $c(G \times G) \subset [G,G]$ and $[G,G]$ is almost Zariski closed, we conclude immediately that $[\overline{\overline{G}}, \overline{\overline{G}}]^\circ \subset [G,G]$ (cf. Exer. 4.3#10). This is almost what we want, but some additional theory (which we omit) is required in order to show that $[\overline{\overline{G}}, \overline{\overline{G}}]$ is connected, rather than having finitely many components.

Because $[\overline{\overline{G}}, \overline{\overline{G}}] \subset G$, it is immediate that $\overline{\overline{G}}/G$ is abelian. $\square$

For connected groups, we now show that tori present the only obstruction to being almost Zariski closed.

(4.6.8) Corollary. *If G is any connected subgroup of $\mathrm{SL}(\ell, \mathbb{R})$, then there is a connected, almost-Zariski closed torus T of $\overline{\overline{G}}$, such that GT is almost Zariski closed.*

Proof. Write $\overline{\overline{G}}^\circ = (TL) \ltimes U$, with T, L, U as in Thm. 4.4.7. Because $L = [L,L] \subset [\overline{\overline{G}}, \overline{\overline{G}}]$, we know $L \subset G$ (see Cor. 4.6.7). Furthermore, because T normalizes G (see Exer. 4.3#12), we may assume $T \subset G$, by replacing G with GT.

Therefore $G = (TL) \ltimes (U \cap G)$ (see Exer. 3). Furthermore, since G is connected, we know that $U \cap G$ is connected, so $U \cap G$ is Zariski closed (see Prop. 4.6.2). Then Prop. 4.6.4 implies that $G = (TL) \ltimes (U \cap G)$ is almost Zariski closed. $\square$

We will make use of the following technical result:

(4.6.9) Lemma. *Show that if*

- *G is an almost-Zariski closed subgroup of $\mathrm{SL}(\ell, \mathbb{R})$,*

- *H and V are connected subgroups of G that are almost Zariski closed, and*

- *$f\colon V \to G$ is a rational function (not necessarily a homomorphism), with $f(e) = e$,*

then the subgroup $\langle H, f(V) \rangle$ is almost Zariski closed.

Plausibility argument. There is no harm in assuming that $G = \overline{\langle f(H), H \rangle}^\circ$, so we wish to show that H and $f(V)$, taken together, generate G. Since $[G,G]H$ is

- almost Zariski closed (see Prop. 4.6.4),

- contained in $\langle H, f(V) \rangle$ (see Cor. 4.6.7), and

- normal in G (because it contains $[G,G]$),

there is no harm in modding it out. Thus, we may assume that G is abelian and that $H = \{e\}$.

Now, using the fact that G is abelian, we have $G = A \times C \times U$, where A is a hyperbolic torus, C is a compact torus, and U is unipotent (see Thm. 4.4.7 and Cor. 4.4.12). Because these are three completely different types of groups, it is not difficult to believe that there are subgroups A_V, C_V, and U_V of A, C, and V, respectively, such that $\langle f(V) \rangle = A_V \times C_V \times U_V$ (cf. Exer. 4).

Now U_V, being connected and unipotent, is Zariski closed (see Prop. 4.6.2). The other two require some argument. $\square$

Exercises for §4.6.

#1. Show that every unipotent real algebraic group is connected and simply connected.
[*Hint:* See proof of (4.6.2).]

#2. Complete the proof of Cor. 4.6.5, by showing that if T is a hyperbolic torus, and $\rho \colon T \to \mathrm{SL}(m, \mathbb{R})$ is a polynomial homomorphism, then $\rho(T)$ is almost Zariski closed.
[*Hint:* Use Cors. 4.5.4 and 4.4.13.]

#3. Show that if G is a subgroup of a semidirect product $A \ltimes B$, and $A \subset G$, then $G = A \ltimes (G \cap B)$. If, in addition, G is connected, show that $G \cap B$ is connected.

#4. Suppose $Q \colon \mathbb{R} \to \mathbb{R}$ is any nonconstant polynomial with $Q(0) = 0$, and define $f \colon \mathbb{R} \to \mathbb{D}_2 \times \mathbb{U}_2 \subset \mathrm{SL}(4, \mathbb{R})$ by

$$f(t) = \begin{bmatrix} 1 + t^2 & 0 & 0 & 0 \\ 0 & 1/(1 + t^2) & 0 & 0 \\ 0 & 0 & 1 & 0 \\ 0 & 0 & Q(t) & 1 \end{bmatrix}.$$

Show $\langle f(\mathbb{R}) \rangle = \mathbb{D}_2 \times \mathbb{U}_2$.

4.7. Borel Density Theorem

The Borel Density Theorem (4.7.1) is a generalization of the important fact that if $\Gamma = \mathrm{SL}(\ell, \mathbb{Z})$, then $\overline{\overline{\Gamma}} = \mathrm{SL}(\ell, \mathbb{R})$ (see Exer. 1). Because the Zariski closure of Γ is all of $\mathrm{SL}(\ell, \mathbb{R})$, we may say that Γ is **Zariski dense** in $\mathrm{SL}(\ell, \mathbb{R})$. That is why this is known as a "density" theorem.

(4.7.1) Proposition (Borel Density Theorem). *If Γ is any lattice in any closed subgroup G of $\mathrm{SL}(\ell, \mathbb{R})$, then the Zariski closure $\overline{\overline{\Gamma}}$ of Γ contains*

1) *every unipotent element of G and*

2) *every hyperbolic element of G.*

We precede the proof with a remark and two lemmas.

(4.7.2) **Remark.**

    1) If G is a compact group, then the trivial subgroup $\Gamma = \{e\}$ is a lattice in G, and $\overline{\overline{\Gamma}} = \{e\}$ does not contain any nontrivial elements of G. This is consistent with Prop. 4.7.1, because nontrivial elements of a compact group are neither unipotent nor hyperbolic (see Cor. 4.4.10).

    2) Although we do not prove this, $\overline{\overline{\Gamma}}$ actually contains every unipotent or hyperbolic element of $\overline{\overline{G}}$, not only those of G.

(4.7.3) **Lemma** (Poincaré Recurrence Theorem). *Let*

- (Ω, d) *be a metric space;*
- $T \colon \Omega \to \Omega$ *be a homeomorphism; and*
- μ *be a T-invariant probability measure on A.*

Then, for almost every $a \in \Omega$, there is a sequence $n_k \to \infty$, such that $T^{n_k} a \to a$.

Proof. Let

$$A_\epsilon = \{\, a \in \Omega \mid \forall m > 0,\ d(T^m a, a) > \epsilon \,\}.$$

It suffices to show $\mu(A_\epsilon) = 0$ for every ϵ.

    Suppose $\mu(A_\epsilon) > 0$. Then we may choose a subset B of A_ϵ, such that $\mu(B) > 0$ and $\operatorname{diam}(B) < \epsilon$. The sets $B, T^{-1}B, T^{-2}B, \ldots$ cannot all be disjoint, because they all have the same measure and $\mu(\Omega) < \infty$. Hence, $T^{-m}B \cap T^{-n}B \neq \varnothing$, for some $m, n \in \mathbb{Z}^+$ with $m > n$. By applying T^n, we may assume $n = 0$. For $a \in T^{-m}B \cap B$, we have $T^m a \in B$ and $a \in B$, so

$$d(T^m a, a) \leq \operatorname{diam}(B) < \epsilon.$$

Since $a \in B \subset A_\epsilon$, this contradicts the definition of A_ϵ.   □

(4.7.4) **Notation.**

- Recall that the ***projective space*** $\mathbb{R}\mathrm{P}^{m-1}$ is, by definition, the set of one-dimensional subspaces of $\mathbb{R}^m$. Alternatively, $\mathbb{R}\mathrm{P}^{m-1}$ can be viewed as the set of equivalence classes of the equivalence relation on $\mathbb{R}^m \smallsetminus \{0\}$ defined by

$$v \sim w \Leftrightarrow v = \alpha w \text{ for some } \alpha \in \mathbb{R} \smallsetminus \{0\}.$$

  From the alternate description, it is easy to see that $\mathbb{R}\mathrm{P}^{m-1}$ is an $(m-1)$-dimensional smooth manifold (see Exer. 3).

- There is a natural action of $\mathrm{SL}(m, \mathbb{R})$ on $\mathbb{R}\mathrm{P}^{m-1}$, defined by $[v]g = [vg]$, where, for each nonzero $v \in \mathbb{R}^m$, we let $[v] = \mathbb{R}v$ be the image of v in $\mathbb{R}\mathrm{P}^{m-1}$.

(4.7.5) **Lemma.** *Assume*

- *g is an element of $\mathrm{SL}(m, \mathbb{R})$ that is either unipotent or hyperbolic,*
- *μ is a probability measure on the projective space $\mathbb{R}\mathrm{P}^{m-1}$, and*
- *μ is invariant under g.*

Then μ is supported on the set of fixed points of g.

Proof. Let v be any nonzero vector in $\mathbb{R}^m$. For definiteness, let us assume g is unipotent. (See Exer. 4 for a replacement of this paragraph in the case where g is hyperbolic.) Letting $T = g - I$, we know that T is nilpotent (because g is unipotent), so there is some integer $r \geq 0$, such that $vT^r \neq 0$, but $vT^{r+1} = 0$. We have

$$vT^r g = (vT^r)(I + T) = vT^r + vT^{r+1} = vT^r + 0 = vT^r,$$

so $[vT^r] \in \mathbb{R}\mathrm{P}^{m-1}$ is a fixed point for g. Also, for $n \in \mathbb{N}$, we have

$$[v]g^n = \left[\sum_{k=0}^{r} \binom{n}{k} vT^k \right] = \left[\binom{n}{r}^{-1} \sum_{k=0}^{r} \binom{n}{k} vT^k \right] \to [vT^r]$$

(because, for $k < r$, we have $\binom{n}{k}/\binom{n}{r} \to 0$ as $n \to \infty$). Thus, $[v]g^n$ converges to a fixed point of g, as $n \to \infty$.

The Poincaré Recurrence Theorem (4.7.3) implies, for μ-almost every $[v] \in \mathbb{R}\mathrm{P}^{m-1}$, that there is a sequence $n_k \to \infty$, such that $[v]g^{n_k} \to [v]$. On the other hand, we know, from the preceding paragraph, that $[v]g^{n_k}$ converges to a fixed point of g. Thus, μ-almost every element of $\mathbb{R}\mathrm{P}^{m-1}$ is a fixed point of g. In other words, μ is supported on the set of fixed points of g, as desired. $\square$

Proof of the Borel Density Theorem (4.7.1). By Chevalley's Theorem (4.5.2), there exist

- a polynomial homomorphism $\rho\colon \mathrm{SL}(\ell, \mathbb{R}) \to \mathrm{SL}(m, \mathbb{R})$, for some m, and
- a vector $v \in \mathbb{R}^m$,

such that $\overline{\overline{\Gamma}} = \{\, g \in \mathrm{SL}(\ell, \mathbb{R}) \mid vg^\rho \in \mathbb{R}v \,\}$. In other words, letting $[v]$ be the image of v in $\mathbb{R}\mathrm{P}^{m-1}$, we have

$$\overline{\overline{\Gamma}} = \{\, g \in \mathrm{SL}(\ell, \mathbb{R}) \mid [v]g^\rho = [v] \,\}. \qquad (4.7.6)$$

Since $\rho(\Gamma)$ fixes $[v]$, the function ρ induces a well-defined map $\overline{\rho}\colon \Gamma \backslash G \to \mathbb{R}\mathrm{P}^{m-1}$:

$$\overline{\rho}(\Gamma g) = [v]g^\rho.$$

Because Γ is a lattice in G, there is a G-invariant probability measure μ_0 on $\Gamma \backslash G$. The map $\overline{\rho}$ pushes this to a probability measure $\mu = \overline{\rho}_* \mu_0$ on $\mathbb{R}\mathrm{P}^{m-1}$, defined by $\mu(A) = \mu_0(\overline{\rho}^{-1}(A))$ for $A \subset \mathbb{R}\mathrm{P}^{m-1}$. Because μ_0 is G-invariant and ρ is a homomorphism, it is easy to see that μ is $\rho(G)$-invariant.

Let g be any element of G that is either unipotent or hyperbolic. From the conclusion of the preceding paragraph, we know that μ is g^ρ-invariant. Since g^ρ is either unipotent or hyperbolic (see Cor. 4.3.6), Lem. 4.7.5 implies that μ is supported on the set of fixed points of g^ρ. Since $[v]$ is obviously in the support of μ (see Exer. 5), we conclude that $[v]$ is fixed by g^ρ; that is, $[v]g^\rho = [v]$. From (4.7.6), we conclude that $g \in \overline{\overline{\Gamma}}$, as desired. □

Exercises for §4.7.

#1. Show (without using the Borel Density Theorem) that the Zariski closure of $\mathrm{SL}(\ell, \mathbb{Z})$ is $\mathrm{SL}(\ell, \mathbb{R})$.
[*Hint:* Let $\Gamma = \mathrm{SL}(\ell, \mathbb{Z})$, and let $H = \overline{\overline{\Gamma}}^{\circ}$. If $g \in \mathrm{SL}(\ell, \mathbb{Q})$, then $g^{-1}\Gamma g$ contains a finite-index subgroup of Γ. Therefore g normalizes H. Because $\mathrm{SL}(\ell, \mathbb{Q})$ is dense in $\mathrm{SL}(n, \mathbb{R})$, this implies that H is a normal subgroup of $\mathrm{SL}(\ell, \mathbb{R})$. Now apply Eg. 4.4.5(1).]

#2. Use the Borel Density Theorem to show that if Γ is any lattice in $\mathrm{SL}(\ell, \mathbb{R})$, then $\overline{\overline{\Gamma}} = \mathrm{SL}(\ell, \mathbb{R})$.
[*Hint:* $\mathrm{SL}(\ell, \mathbb{R})$ is generated by its unipotent elements.]

#3. Show that there is a natural covering map from the $(m-1)$-sphere S^{m-1} onto $\mathbb{R}\mathrm{P}^{m-1}$, so $\mathbb{R}\mathrm{P}^{m-1}$ is a C^∞ manifold.

#4. In the notation of Lem. 4.7.5, show that if g is hyperbolic, and v is any nonzero vector in $\mathbb{R}^m$, then $[v]g^n$ converges to a fixed point of g, as $n \to \infty$.
[*Hint:* Assume g is diagonal. For $v = (v_1, \ldots, v_m)$, calculate vg^n.]

#5. In the notation of the proof of Prop. 4.7.1, show that the support of μ is the closure of $[v]G^\rho$.
[*Hint:* If some point of $[v]G^\rho$ is contained in an open set of measure 0, then, because μ is invariant under $\rho(G)$, all of $[v]G^\rho$ is contained in an open set of measure 0.]

#6. (*The Borel Density Theorem, essentially as stated by Borel*) Suppose
- G is a connected, semisimple subgroup of $\mathrm{SL}(\ell, \mathbb{R})$, such that every simple factor of G is noncompact, and
- Γ is a lattice in G.

Show:

(a) $G \subset \overline{\overline{\Gamma}}$,

(b) Γ is not contained in any proper, closed subgroup of G that has only finitely many connected components, and

(c) if $\rho \colon G \to \mathrm{SL}(m, \mathbb{R})$ is any continuous homomorphism, then every element of $\rho(G)$ is a finite linear combination (with real coefficients) of elements of $\rho(\Gamma)$.

[*Hint:* Use Prop. 4.7.1. (6c) The subspace of $\mathrm{Mat}_{m \times m}(\mathbb{R})$ spanned by $\rho(\Gamma)$ is invariant under multiplication by $\rho(\Gamma)$, so it must be invariant under multiplication by $\rho(G)$.]

#7. Suppose
- G is a closed, connected subgroup of $\mathrm{SL}(\ell, \mathbb{R})$, and
- Γ is a lattice in G.

Show there are only countably many closed, connected subgroups S of G, such that

(a) $\Gamma \cap S$ is a lattice in S, and

(b) there is a one-parameter unipotent subgroup u^t of S, such that $(\Gamma \cap S)\{u^t\}$ is dense in S.

[*Hint:* You may assume, without proof, the fact that every lattice in every connected Lie group is finitely generated. Show $S \subset \overline{\overline{\Gamma \cap S}}$. Conclude that S is uniquely determined by $\Gamma \cap S$.]

#8. Suppose
- G is an almost-Zariski closed subgroup of $\mathrm{SL}(\ell, \mathbb{R})$,
- U is a connected, unipotent subgroup of G,
- Γ is a discrete subgroup of G,
- μ is an ergodic U-invariant probability measure on $\Gamma \backslash G$, and
- there does **not** exist a subgroup H of G, such that
    - H is almost Zariski closed,
    - $U \subset H$, and
    - some H-orbit has full measure.

Show, for all $x \in \Gamma \backslash G$ and every subset V of G, that if $\mu(xV) > 0$, then $G \subset \overline{\overline{V}}$.

[*Hint:* Assume V is Zariski closed and irreducible. Let

$U_{xV} = \{ u \in U \mid xVu = xV \}$ and $U_V = \{ u \in U \mid Vu = V \}$.

Assuming that V is minimal with $\mu(xV) > 0$, we have

$$\mu(xV \cap xVu) = 0 \text{ for } u \in U \smallsetminus U_{xV}.$$

So U/U_{xV} is finite. Since U is connected, then $U_{xV} = U$. Similarly (and because Γ is countable), U/U_V is countable, so $U_V = U$.

Let $\Gamma_V = \{ y \in \Gamma \mid Vy = V \}$. Then μ defines a measure μ_V on $\Gamma_V \backslash V$, and this pushes to a measure $\overline{\mu_V}$ on $\overline{\Gamma_V} \backslash V$. By combining Chevalley's Theorem (4.5.2), the Borel Density Theorem (4.7.5), and the ergodicity of U, conclude that $\overline{\mu_V}$ is supported on a single point. Letting $H = \langle \overline{\Gamma_V}, U \rangle$, some H-orbit has positive measure, and is contained in xV.]

4.8. Subgroups defined over $\mathbb{Q}$

In this section, we briefly discuss the relationship between lattice subgroups and the integer points of a group. This material is not needed for the proof of Ratner's Theorem, but it is related, and it is used in many applications, including Margulis' Theorem on values of quadratic forms (1.2.2).

(4.8.1) Definition. A Zariski closed subset Z of $\mathrm{SL}(\ell, \mathbb{R})$ is said to be **defined over** $\mathbb{Q}$ if the defining polynomials for Z can be taken to have all of their coefficients in $\mathbb{Q}$; that is, if $Z = \mathrm{Var}(Q)$ for some subset Q of $\mathbb{Q}[x_{1,1}, \ldots, x_{\ell,\ell}]$.

(4.8.2) Example. The algebraic groups in (1–5) of Eg. 4.1.2 are defined over $\mathbb{Q}$. Those in (6–8) may or may not be defined over $\mathbb{Q}$, depending on the particular choice of v, V, or Q. Namely:

A) The stabilizer of a vector v is defined over $\mathbb{Q}$ if and only if v is a scalar multiple of a vector in $\mathbb{Z}^\ell$ (see Exer. 3).

B) The stabilizer of a subspace V of $\mathbb{R}^\ell$ is defined over $\mathbb{Q}$ if and only if V is spanned by vectors in $\mathbb{Z}^\ell$ (see Exer. 4).

C) The special orthogonal group $\mathrm{SO}(Q)$ of a nondegenerate quadratic form Q is defined over $\mathbb{Q}$ if and only if Q is a scalar multiple of a form with integer coefficients (see Exer. 5).

(4.8.3) Definition. A polynomial function $\phi \colon H \to \mathrm{SL}(n, \mathbb{R})$ is **defined over** $\mathbb{Q}$ if it can be obtained as in Defn. 4.3.5, but with $\mathbb{R}[x_{1,1}, \ldots, x_{\ell,\ell}]$ replaced by $\mathbb{Q}[x_{1,1}, \ldots, x_{\ell,\ell}]$ in 4.3.5(1). That is, only polynomials with rational coefficients are allowed in the construction of ϕ.

The fact that $\mathbb{Z}^k$ is a lattice in $\mathbb{R}^k$ has a vast generalization:

(4.8.4) Theorem (Borel and Harish-Chandra). *Suppose*

- *G is a Zariski closed subgroup of $\mathrm{SL}(\ell, \mathbb{R})$,*
- *G is defined over $\mathbb{Q}$, and*
- *no nontrivial polynomial homomorphism from G° to $\mathbb{D}_2$ is defined over $\mathbb{Q}$,*

then $G \cap SL(\ell, \mathbb{Z})$ is a lattice in G.

(4.8.5) Corollary. $SL(\ell, \mathbb{Z})$ *is a lattice in* $SL(\ell, \mathbb{R})$.

(4.8.6) Example. $\mathbb{D}_2 \cap SL(2, \mathbb{Z}) = \{\pm I\}$ is finite, so it is *not* a lattice in $\mathbb{D}_2$.

It is interesting to note that Cor. 4.8.5 can be proved from properties of unipotent flows. (One can then use this to obtain the general case of Thm. 4.8.4, but this requires some of the theory of "arithmetic groups" (cf. Exer. 11).)

Direct proof of Cor. 4.8.5. Let $G = SL(\ell, \mathbb{R})$ and $\Gamma = SL(\ell, \mathbb{Z})$. For

- a nontrivial, unipotent one-parameter subgroup u^t, and
- a compact subset K of $\Gamma \backslash G$,

we define $f : \Gamma \backslash G \to \mathbb{R}^{\geq 0}$ by

$$f(x) = \liminf_{L \to \infty} \frac{1}{L} \int_0^L \chi_K(xu^t)\, dt,$$

where χ_K is the characteristic function of K.

The key to the proof is that the conclusion of Thm. 1.9.2 can be proved by using the polynomial nature of u^t — *without* knowing that Γ is a lattice. Furthermore, a single compact set K can be chosen to work for all x in any compact subset of $\Gamma \backslash G$. This means that, by choosing K appropriately, we may assume that $f > 0$ on some nonempty open set.

Letting μ be the Haar measure on $\Gamma \backslash G$, we have $\int_{\Gamma \backslash G} f\, d\mu \leq \mu(K) < \infty$, so $f \in L^1(\Gamma \backslash G, \mu)$.

It is easy to see, from the definition, that f is u^t-invariant. Therefore, the Moore Ergodicity Theorem implies that f is essentially G-invariant (see Exer. 3.2#6). So f is essentially constant.

If a nonzero constant is in L^1, then the space must have finite measure. So Γ is a lattice. $\qquad \square$

Exercises for §4.8.

#1. Show that if C is any subset of $SL(\ell, \mathbb{Q})$, then $\overline{\overline{C}}$ is defined over $\mathbb{Q}$.
   [*Hint:* Suppose $\overline{\overline{C}} = \text{Var}(\mathcal{Q})$, for some $\mathcal{Q} \subset \mathcal{P}^d$, where $\mathcal{P}^d$ is the set of polynomials of degree $\leq d$. Because the subspace $\{ Q \in \mathcal{P}^d \mid Q(C) = 0 \}$ of $\mathcal{P}^d$ is defined by linear equations with rational coefficients, it is spanned by rational vectors.]

#2. (*Requires some commutative algebra*) Let Z be a Zariski closed subset of $SL(\ell, \mathbb{R})$. Show that Z is defined over $\mathbb{Q}$ if and only if $\sigma(Z) = Z$, for every Galois automorphism ϕ of $\mathbb{C}$.

[*Hint:* ($\Leftarrow$) You may assume Hilbert's Nullstellensatz, which implies there is a subset Q of $\overline{Q}[x_{1,1}, \ldots, x_{\ell,\ell}]$, such that $\overline{\overline{C}} = \text{Var}(Q)$, where $\overline{Q}$ is the algebraic closure of Q. Then $\overline{Q}$ may be replaced with some finite Galois extension F of $\mathbb{Q}$, with Galois group Φ. For $Q \in \mathcal{Q}$, any symmetric function of $\{ Q^{\phi} \mid \phi \in \Phi \}$ has rational coefficients.]

#3. Verify Eg. 4.8.2(A).

[*Hint:* ($\Rightarrow$) The vector v fixed by $\text{Stab}_{\text{SL}(\ell, \mathbb{R})}(v)$ is unique, up to a scalar multiple. Thus, $v^{\phi} \in \mathbb{R}v$, for every Galois automorphism ϕ of $\mathbb{C}$. Assuming some coordinate of v is rational (and nonzero), then all the coordinates of v must be rational.]

#4. Verify Eg. 4.8.2(B).

[*Hint:* ($\Rightarrow$) Cf. Hint to Exer. 3. Any nonzero vector in V with the minimal number of nonzero coordinates (and some coordinate rational) must be fixed by each Galois automorphism of $\mathbb{C}$. So V contains a rational vector v. By a similar argument, there is a rational vector that is linearly independent from v. By induction, create a basis of rational vectors.]

#5. Verify Eg. 4.8.2(C).

[*Hint:* ($\Rightarrow$) Cf. Hint to Exer. 3. The quadratic form Q is unique, up to a scalar multiple.]

#6. Suppose Q is a quadratic form on $\mathbb{R}^n$, such that $\overline{\overline{\text{SO}(Q)^{\circ}}}$ is defined over $\mathbb{Q}$. Show that Q is a scalar multiple of a form with integer coefficients.

[*Hint:* The invariant form corresponding to $\text{SO}(Q)$ is unique up to a scalar multiple. We may assume one coefficient is 1, so Q is fixed by every Galois automorphism of $\mathbb{C}$.]

#7. Suppose

- G is a Zariski closed subgroup of $\text{SL}(\ell, \mathbb{R})$,

- G° is generated by its unipotent elements, and

- $G \cap \text{SL}(\ell, \mathbb{Z})$ is a lattice in G.

Show G is defined over $\mathbb{Q}$.

[*Hint:* Use the Borel Density Theorem (4.7.1).]

#8. Suppose $\sigma \colon G \to \text{SL}(m, \mathbb{R})$ is a polynomial homomorphism that is defined over $\mathbb{Q}$. Show:

(a) $\sigma(G \cap \text{SL}(\ell, \mathbb{Z})) \subset \text{SL}(m, \mathbb{Q})$, and

(b) there is a finite-index subgroup Γ of $G \cap \text{SL}(\ell, \mathbb{Z})$, such that $\sigma(\Gamma) \subset \text{SL}(m, \mathbb{Z})$.

[*Hint:* (8b) There is a nonzero integer k, such that if $g \in G \cap \text{SL}(\ell, \mathbb{Z})$ and $g \equiv I \pmod{k}$, then $\sigma(g) \in \text{SL}(m, \mathbb{Z})$.]

#9. Suppose G is a Zariski closed subgroup of $SL(\ell, \mathbb{R})$. Show that if some nontrivial polynomial homomorphism from G° to $\mathbb{D}_2$ is defined over $\mathbb{Q}$, then $G \cap SL(\ell, \mathbb{Z})$ is **not** a lattice in G.

#10. Show that if G is a Zariski closed subgroup of $SL(\ell, \mathbb{R})$ that is defined over $\mathbb{Q}$, and G° is generated by its unipotent elements, then $G \cap SL(\ell, \mathbb{Z})$ is a lattice in G.

#11. Suppose
- G is a connected, noncompact, simple subgroup of $SL(\ell, \mathbb{R})$,
- $\Gamma = G \cap SL(\ell, \mathbb{Z})$, and
- the natural inclusion $\tau : \Gamma \backslash G \hookrightarrow SL(\ell, \mathbb{Z}) \backslash SL(\ell, \mathbb{R})$, defined by $\tau(\Gamma x) = SL(\ell, \mathbb{Z})x$, is proper.

Show (without using Thm. 4.8.4) that Γ is a lattice in G. [*Hint:* See the proof of Cor. 4.8.5.]

4.9. Appendix on Lie groups

In this section, we briefly recall (without proof) some facts from the theory of Lie groups.

(4.9.1) Definition. A group G is a *Lie group* if the underlying set is a C^∞ manifold, and the group operations (multiplication and inversion) are C^∞ functions.

A closed subset of a Lie group need not be a manifold (it could be a Cantor set, for example), but this phenomenon does not occur for subgroups:

(4.9.2) Theorem. *Any closed subgroup of a Lie group is a Lie group.*

It is easy to see that the universal cover of a (connected) Lie group is a Lie group.

(4.9.3) Definition. Two connected Lie groups G and H are *locally isomorphic* if their universal covers are C^∞ isomorphic.

We consider only *linear* Lie groups; that is, Lie groups that are closed subgroups of $SL(\ell, \mathbb{R})$, for some ℓ. The following classical theorem shows that, up to local isomorphism, this results in no loss of generality.

(4.9.4) Theorem (Ado-Iwasawa). *Any connected Lie group is locally isomorphic to a closed subgroup of $SL(\ell, \mathbb{R})$, for some ℓ.*

It is useful to consider subgroups that need not be closed, but may only be immersed submanifolds:

(4.9.5) **Definition.** A subgroup H of a Lie group G is a **Lie subgroup** if there is a Lie group H_0 and an injective C^∞ homomorphism $\sigma: H_0 \to G$, such that $H = \sigma(H_0)$. Then we consider H to be a Lie group, by giving it a topology that makes σ a homeomorphism. (If H is not closed, this is **not** the topology that H acquires by being a subset of G.)

(4.9.6) **Definition.** Let G be a Lie subgroup of $\mathrm{SL}(\ell, \mathbb{R})$. The tangent space to G at the identity element e is the **Lie algebra** of G. It is, by definition, a vector subspace of the space $\mathrm{Mat}_{\ell \times \ell}(\mathbb{R})$ of $\ell \times \ell$ real matrices.

  The Lie algebra of a Lie group G, H, U, S, etc., is usually denoted by the corresponding lower-case gothic letter $\mathfrak{g}$, $\mathfrak{h}$, $\mathfrak{u}$, $\mathfrak{s}$, etc.

(4.9.7) **Example.**

  1) The Lie algebra of $\mathbb{U}_\ell$ is

$$\mathfrak{U}_\ell = \begin{bmatrix} 0 & & \mathbf{0} \\ & \ddots & \\ \mathbf{*} & & 0 \end{bmatrix},$$

  the space of strictly lower-triangular matrices.

  2) Let $d\det: \mathrm{Mat}_{n \times n}(\mathbb{R}) \to \mathbb{R}$ be the derivative of the determinant map det at the identity matrix I. Then $(d\det)(A) = \mathrm{trace}\, A$. Therefore the Lie algebra of $\mathrm{SL}(\ell, \mathbb{R})$ is

$$\mathfrak{sl}(\ell, \mathbb{R}) = \{ A \in \mathrm{Mat}_{\ell \times \ell}(\mathbb{R}) \mid \mathrm{trace}\, A = 0 \}.$$

  So the Lie algebra of any Lie subgroup of $\mathrm{SL}(\ell, \mathbb{R})$ is contained in $\mathfrak{sl}(\ell, \mathbb{R})$.

  3) The Lie algebra of $\mathbb{D}_\ell$ is

$$\mathfrak{D}_\ell = \left\{ \begin{bmatrix} a_1 & & \mathbf{0} \\ & \ddots & \\ \mathbf{0} & & a_\ell \end{bmatrix} \,\middle|\, a_1 + \cdots + a_\ell = 0 \right\},$$

  the space of diagonal matrices of trace 0.

(4.9.8) **Definition.**

  1) For $\underline{x}, \underline{y} \in \mathrm{Mat}_{\ell \times \ell}(\mathbb{R})$, let $[\underline{x}, \underline{y}] = \underline{x}\underline{y} - \underline{y}\underline{x}$. This is the **Lie bracket** of $\underline{x}$ and $\underline{y}$.

  2) A vector subspace $\mathfrak{h}$ of $\mathrm{Mat}_{\ell \times \ell}(\mathbb{R})$ is a **Lie subalgebra** if $[\underline{x}, \underline{y}] \in V$ for all $\underline{x}, \underline{y} \in \mathfrak{h}$.

  3) A linear map $\tau: \mathfrak{g} \to \mathfrak{h}$ between Lie subalgebras is a **Lie algebra homomorphism** if $\tau([\underline{x}, \underline{y}]) = [\tau(\underline{x}), \tau(\underline{y})]$ for all $\underline{x}, \underline{y} \in \mathfrak{g}$.

(4.9.9) Proposition.

1) *The Lie algebra of any Lie subgroup of* $\mathrm{SL}(\ell, \mathbb{R})$ *is a Lie subalgebra.*

2) *Any Lie subalgebra* $\mathfrak{h}$ *of* $\mathfrak{sl}(\ell, \mathbb{R})$ *is the Lie algebra of a unique* **connected** *Lie subgroup H of* $\mathrm{SL}(\ell, \mathbb{R})$.

3) *The differential of a Lie group homomorphism is a Lie algebra homomorphism. That is, if $\phi\colon G \to H$ is a C^∞ Lie group homomorphism, and $D\phi$ is the derivative of ϕ at e, then $D\phi$ is a Lie algebra homomorphism from $\mathfrak{g}$ to $\mathfrak{h}$.*

4) *A connected Lie group is uniquely determined (up to local isomorphism) by its Lie algebra. That is, two connected Lie groups G and H are locally isomorphic if and only if their Lie algebras are isomorphic.*

(4.9.10) Definition. The *exponential map*

$$\exp\colon \mathfrak{sl}(\ell, \mathbb{R}) \to \mathrm{SL}(\ell, \mathbb{R})$$

is defined by the usual power series

$$\exp \underline{x} = I + \underline{x} + \frac{\underline{x}^2}{2!} + \frac{\underline{x}^3}{3!} + \cdots$$

(4.9.11) Example. Let

$$\underline{a} = \begin{bmatrix} 1 & 0 \\ 0 & -1 \end{bmatrix}, \underline{u} = \begin{bmatrix} 0 & 0 \\ 1 & 0 \end{bmatrix}, \text{ and } \underline{v} = \begin{bmatrix} 0 & 1 \\ 0 & 0 \end{bmatrix}.$$

Then, letting

$$a^s = \begin{bmatrix} e^s & 0 \\ 0 & e^{-s} \end{bmatrix}, u^t = \begin{bmatrix} 1 & 0 \\ t & 1 \end{bmatrix}, v^r = \begin{bmatrix} 1 & r \\ 0 & 1 \end{bmatrix},$$

as usual in $\mathrm{SL}(2, \mathbb{R})$, it is easy to see that:

1) $\exp(s\underline{a}) = a^s$,
2) $\exp(t\underline{u}) = u^t$,
3) $\exp(r\underline{v}) = v^r$,
4) $[\underline{u}, \underline{a}] = 2\underline{u}$,
5) $[\underline{v}, \underline{a}] = -2\underline{v}$, and
6) $[\underline{v}, \underline{u}] = \underline{a}$.

(4.9.12) Proposition. *Let $\mathfrak{g}$ be the Lie algebra of a Lie subgroup G of* $\mathrm{SL}(\ell, \mathbb{R})$. *Then:*

1) $\exp \mathfrak{g} \subset G$.

2) *For any $\underline{g} \in \mathfrak{g}$, the map $\mathbb{R} \to G$ defined by $g^t = \exp(t\underline{g})$ is a one-parameter subgroup of G.*

3) *The restriction of* exp *to some neighborhood of* 0 *in* $\mathfrak{g}$ *is a diffeomorphism onto some neighborhood of* e *in* G.

(4.9.13) Definition.

1) A group G is *solvable* if there is a chain
$$e = G_0 \lhd G_1 \lhd \cdots \lhd G_k = G$$
of subgroups of G, such that, for $1 \le i \le k$,
   (a) G_{i-1} is a normal subgroup of G_i, and
   (b) the quotient group G_i/G_{i-1} is abelian.

2) Any Lie group G has a unique maximal closed, connected, solvable, normal subgroup. This is called the *radical* of G, and is denoted Rad G.

3) A Lie group G is said to be *semisimple* if Rad $G = \{e\}$.

(4.9.14) Remark. According to Defn. 4.4.1, G is semisimple if G° has no nontrivial, connected, *abelian*, normal subgroups. One can show that this implies there are also no nontrivial, connected, *solvable* normal subgroups.

(4.9.15) Theorem. *Any Lie group G has a closed, semisimple subgroup L, such that*

1) *L is semisimple and*

2) *$G = L$ Rad G.*

*The subgroup L is called a **Levi subgroup** of G; it is usually not unique.*

(4.9.16) Warning. The above definition is from the theory of Lie groups. In the theory of algebraic groups, the term *Levi subgroup* is usually used to refer to a slightly different subgroup — namely, the subgroup LT of Thm. 4.4.7.

(4.9.17) Theorem (Lie-Kolchin Theorem). *If G is any connected, solvable Lie subgroup of* $\mathrm{SL}(\ell, \mathbb{R})$, *then there exists $h \in \mathrm{SL}(\ell, \mathbb{C})$, such that $h^{-1}Gh \subset \mathbb{D}_\ell \mathbb{U}_\ell$.*

(4.9.18) Definition. Let $\mathfrak{g}$ be the Lie algebra of a Lie subgroup G of $\mathrm{SL}(\ell, \mathbb{R})$.

- We use $\mathrm{GL}(\mathfrak{g})$ to denote the group of all invertible linear transformations $\mathfrak{g} \to \mathfrak{g}$. This is a Lie group, and its Lie algebra $\mathfrak{gl}(\mathfrak{g})$ consists of all (not necessarily invertible) linear transformations $\mathfrak{g} \to \mathfrak{g}$.

- We define a group homomorphism $\mathrm{Ad}_G \colon G \to \mathrm{GL}(\mathfrak{g})$ by
$$\underline{x}(\mathrm{Ad}_G g) = g^{-1}\underline{x}g.$$

Note that $\text{Ad}_G\,g$ is the derivative at e of the group automorphism $x \mapsto g^{-1}xg$, so Ad_G is a Lie algebra automorphism.

- We define a Lie algebra homomorphism $\text{ad}_{\mathfrak{g}} : \mathfrak{g} \to \mathfrak{gl}(\mathfrak{g})$ by

$$\underline{x}(\text{ad}_{\mathfrak{g}}\,\underline{g}) = [\underline{x},\underline{g}].$$

We remark that $\text{ad}_{\mathfrak{g}}$ is the derivative at e of Ad_G.

(4.9.19) Remark. A Lie group G is unimodular (that is, the right Haar measure is also invariant under left translations) if and only if $\det(\text{Ad}_G\,g) = 1$, for all $g \in G$.

(4.9.20) Proposition. *The maps* exp, Ad_G *and* $\text{ad}_{\mathfrak{g}}$ *are natural. That is, if* $\rho : G \to H$ *is a Lie group homomorphism, and* $d\rho$ *is the derivative of* ρ *at* e, *then*

1) $(\exp \underline{g})^{\rho} = \exp(d\rho(\underline{g}))$,

2) $d\rho(\underline{x}(\text{Ad}_G\,g)) = (d\rho\underline{x})(\text{Ad}_H\,g^{\rho})$, *and*

3) $d\rho(\underline{x}(\text{ad}_{\mathfrak{g}}\,\underline{g})) = (d\rho\underline{x})(\text{Ad}_{\mathfrak{h}}\,d\rho(\underline{g}))$.

(4.9.21) Corollary. *We have* $\text{Ad}_G(\exp \underline{g}) = \exp(\text{ad}_{\mathfrak{g}}\,\underline{g})$. *That is,*

$$\underline{x}(\text{Ad}_G(\exp \underline{g})) = \underline{x} + \underline{x}(\text{ad}_{\mathfrak{g}}\,\underline{g}) + \frac{1}{2}\underline{x}(\text{ad}_{\mathfrak{g}}\,\underline{g})^2 + \frac{1}{3!}\underline{x}(\text{ad}_{\mathfrak{g}}\,\underline{g})^3 + \cdots$$

The commutation relations (4,5,6) of Eg. 4.9.11 lead to a complete understanding of all $\mathfrak{sl}(2,\mathbb{R})$-modules:

(4.9.22) Proposition. *Suppose*

- $\mathcal{W}$ *is a finite-dimensional real vector space, and*

- $\rho : \mathfrak{sl}(2,\mathbb{R}) \to \mathfrak{sl}(\mathcal{W})$ *is a Lie algebra homomorphism.*

Then there is a sequence $\lambda_1,\dots,\lambda_n$ *of natural numbers, and a basis*

$$\left\{ w_{i,j} \;\middle|\; \begin{matrix} 1 \le i \le n, \\ 0 \le j \le \lambda_i \end{matrix} \right\}$$

of $\mathcal{W}$, *such that, for all* i,j, *we have:*

1) $w_{i,j}\underline{a}^{\rho} = (2j - \lambda_i)w_{i,j}$,

2) $w_{i,j}\underline{u}^{\rho} = (\lambda_i - j)w_{i,j+1}$, *and*

3) $w_{i,j}\underline{v}^{\rho} = jw_{i,j-1}$.

(4.9.23) Remark. The above proposition has the following immediate consequences.

1) Each $w_{i,j}$ is an eigenvector for $\underline{a}^{\rho}$, and all of the eigenvalues are integers. (Therefore, $\underline{a}^{\rho}$ is diagonalizable over $\mathbb{R}$.)

2) For any integer λ, we let

$$\mathcal{W}_\lambda = \{\, w \in \mathcal{W} \mid w\underline{a}^\rho = \lambda w \,\}.$$

This is called the **weight space** corresponding to λ. (If λ is an eigenvalue, it is the corresponding eigenspace; otherwise, it is $\{0\}$.) A basis of $\mathcal{W}_\lambda$ is given by

$$\left\{\, w_{i,(\lambda+\lambda_i)/2} \;\middle|\; \begin{array}{l} 1 \le i \le n, \\ \lambda_i \le |\lambda| \end{array} \,\right\}.$$

3) For all λ, we have $\mathcal{W}_\lambda \underline{u}^\rho \subset \mathcal{W}_{\lambda+2}$ and $\mathcal{W}_\lambda \underline{v}^\rho \subset \mathcal{W}_{\lambda-2}$.

4) The kernel of $\underline{u}^\rho$ is spanned by $\{w_{1,\lambda_1}, \ldots, w_{n,\lambda_n}\}$, and the kernel of $\underline{v}^\rho$ is spanned by $\{w_{1,0}, \ldots, w_{n,0}\}$.

5) $\underline{u}^\rho$ and $\underline{v}^\rho$ are nilpotent.

Exercises for §4.9.

#1. Suppose u^t is a nontrivial, one-parameter, unipotent subgroup of $\mathrm{SL}(2, \mathbb{R})$.

    (a) Show that $\{u^t\}$ is conjugate to $\mathbb{U}_2$.

    (b) Suppose a^s is a nontrivial, one-parameter, hyperbolic subgroup of $\mathrm{SL}(2, \mathbb{R})$ that normalizes $\{u^t\}$. Show there exists $h \in \mathrm{SL}(2, \mathbb{R})$, such that $h^{-1}\{a^s\}h = \mathbb{D}_2$ and $h^{-1}\{u^t\}h = \mathbb{U}_2$.

#2. Verify the calculations of Eg. 4.9.11.

#3. Show that if a is a hyperbolic element of $\mathrm{SL}(\ell, \mathbb{R})$, and V is an a-invariant subspace of $\mathbb{R}^\ell$, then V has an a-invariant complement. That is, there is an a-invariant subspace W of $\mathbb{R}^\ell$, such that $\mathbb{R}^\ell = V \oplus W$.
[*Hint:* A subspace of $\mathbb{R}^\ell$ is a-invariant if and only if it is a sum of subspaces of eigenspaces.]

#4. Show that if L is a Levi subgroup of G, then $L \cap \mathrm{Rad}\, G$ is discrete.
[*Hint:* $L \cap \mathrm{Rad}\, G$ is a closed, solvable, normal subgroup of L.]

#5. Show that every continuous homomorphism $\rho \colon \mathbb{R}^k \to \mathbb{R}$ is a linear map.
[*Hint:* Every homomorphism is $\mathbb{Q}$-linear. Use continuity to show that ρ is $\mathbb{R}$-linear.]

#6. Suppose T is a connected Lie subgroup of $\mathbb{D}_\ell$, and $\rho \colon T \to \mathbb{R}^+$ is a C^∞ homomorphism.

    (a) Show there exist real numbers $\alpha_1, \cdots, \alpha_\ell$, such that

$$\rho(t) = t_{1,1}^{\alpha_1} \cdots t_{\ell,\ell}^{\alpha_\ell}$$

for all $t \in T$.

(b) Show that if ρ is polynomial, then $\alpha_1, \cdots, \alpha_\ell$ are integers.

[*Hint:* (6a) Use Exer. 5.]

#7. Suppose $\mathcal{W}$ and ρ are as in Prop. 4.9.22. Show:

    (a) No proper $\rho(\mathfrak{sl}(2, \mathbb{R}))$-invariant subspace of $\mathcal{W}$ contains $\ker \underline{u}^\rho$.

    (b) If V and W are $\rho(\mathfrak{sl}(2, \mathbb{R}))$-invariant subspaces of $\mathcal{W}$, such that $V \subsetneq W$, then $V \cap \ker \underline{u}^\rho \subsetneq W \cap \ker \underline{u}^\rho$.

[*Hint:* (7b) Apply (7a) with W in the place of $\mathcal{W}$.]

#8. Suppose
$$\left\{ w_{i,j} \;\middle|\; \begin{array}{l} 1 \le i \le n, \\ 0 \le j \le \lambda_i \end{array} \right\}$$
is a basis of a real vector space $\mathcal{W}$, for some sequence $\lambda_1, \ldots, \lambda_n$ of natural numbers. Show that the equations 4.9.22(1,2,3) determine linear transformations $\underline{a}^\rho$, u^ρ, and $\underline{v}^\rho$ on $\mathcal{W}$, such that the commutation relations (4,5,6) of Eg. 4.9.11 are satisfied. Thus, there is a Lie algebra homomorphism $\sigma \colon \mathfrak{sl}(2, \mathbb{R}) \to \mathfrak{sl}(\mathcal{W})$, such that $\sigma(\underline{a}) = \underline{a}^\rho$, $\sigma(\underline{u}) = \underline{u}^\rho$, and $\sigma(\underline{v}) = \underline{v}^\rho$.

Notes

The algebraic groups that appear in these lectures are defined over $\mathbb{R}$, and our only interest is in their real points. Furthermore, we are interested only in linear groups (that is, subgroups of $\mathrm{SL}(\ell, \mathbb{R})$), not "abelian varieties." Thus, our definitions and terminology are tailored to this setting.

There are many excellent textbooks on the theory of (linear) algebraic groups, including [5, 16], but they generally focus on algebraic groups over $\mathbb{C}$ (or some other algebraically closed field). The books of V. Platonov and A. Rapinchuk [23, Chap. 3] and A. L. Onishchik and E. B. Vinberg [22] are excellent sources for information on algebraic groups over $\mathbb{R}$.

§4.1. Standard textbooks discuss varieties, Zariski closed sets, algebraic groups, dimension, and the singular set.

Whitney's Theorem (4.1.3) appears in [23, Cor. 1 of Thm. 3.6, p. 121].

§4.2. The Zariski closure is a standard topic.

The notion of being "almost Zariski closed" does not arise over an algebraically closed field, so it is not described in most texts. Relevant material (though without using this terminology) appears in [23, §3.2] and [29, §3.1]. References to numerous specific results on almost-Zariski closed subgroups can be found in [28, §3].

Exercise 4.2#2 is a version of [16, Prop. 8.2b, p. 59].

§4.3. Polynomials, unipotent elements, and the Jordan decomposition are standard material. However, most texts consider the Jordan decomposition over $\mathbb{C}$, not $\mathbb{R}$. (Hyperbolic elements and elliptic elements are lumped together into a single class of "semisimple" elements.)

The real Jordan decomposition appears in [11, Lem. IX.7.1, p. 430], for example.

A solution of Exer. 4.3#12 appears in the proof of [29, Thm. 3.2.5, p. 42].

§4.4. This material is standard, except for Thm. 4.4.9 and its corollary (which do not occur over an algebraically closed field).

The theory of roots and weights is described in many textbooks, including [11, 15, 25]. See [11, Table V, p. 518] for a list of the almost-simple groups.

For the case of semisimple groups, the difficult direction ($\Leftarrow$) of Thm. 4.4.9 is immediate from the "Iwasawa decomposition" $G = KAN$, where K is compact, A is a hyperbolic torus, and N is unipotent. This decomposition appears in [23, Thm. 3.9, p. 131], or in many texts on Lie groups.

The proof of Engel's Theorem in Exer. 4.4#5 is taken from [16, Thm. 17.5, p. 112]. The theorem of Burnside mentioned there (or the more general Jacobson Density Theorem) appears in graduate algebra texts, such as [17, Cor. 3.4 of Chap. XVII].

§4.5. This is standard.

§4.6. These results are well known, but do not appear in most texts on algebraic groups.

Proposition 4.6.1 is due to C. Chevalley [9, Prop. 2, §VI.5.2, p. 230]. A proof also appears in [1, §8.6].

See [13, Thm. 8.1.1, p. 107] for a proof of Prop. 4.6.2.

See [13, Thm. 8.3.2, p. 112] for a proof of Thm. 4.6.3.

The analogue of Prop. 4.6.4 over an algebraically closed field is standard (e.g., [16, Cor. 7.4, p. 54]). For a derivation of Prop. 4.6.4 from this, see [28, Lem. 3.17].

See [23, Cor. 1 of Prop. 3.3, p. 113] for a proof of Cor. 4.6.5.

Corollary 4.6.6 is proved in [8, Thm. 15, §II.14, p. 177] and [13, Thm 8.3.3, p. 113].

Completing the proof of Cor. 4.6.7 requires one to know that $\overline{\overline{G}}/G$ is abelian for every connected, semisimple subgroup G of

$SL(\ell, \mathbb{R})$. In fact, $\overline{\overline{G}}/G$ is trivial if $\overline{\overline{G}}$ is "simply connected" as an algebraic group [23, Prop. 7.6, p. 407], and the general case follows from this by using an exact sequence of Galois cohomology groups: $\widetilde{G}_{\mathbb{R}} \to (\widetilde{G}/Z)_{\mathbb{R}} \to \mathcal{H}^1(\mathbb{C}/\mathbb{R}, Z_{\mathbb{C}})$.

A proof of Lem. 4.6.9 appears in [7, §2.2]. (It is based on the analogous result over an algebraically closed field, which is a standard result that appears in [16, Prop. 7.5, p. 55], for example.)

Exercise 4.6#1 is a version of [12, Thm. 8.1.1, p. 107].

§4.7. This material is fairly standard in ergodic theory, but not common in texts on algebraic groups.

The Borel Density Theorem (4.7.1) was proved for semisimple groups in [3] (see Exer. 4.7#6). (The theorem also appears in [19, Lem. II.2.3 and Cor. II.2.6, p. 84] and [29, Thm. 3.2.5, pp. 41–42], for example.) The generalization to all Lie groups is due to S. G. Dani [10, Cor. 2.6].

The Poincaré Recurrence Theorem (4.7.3) can be found in many textbooks on ergodic theory, including [2, Cor. I.1.8, p. 8].

See [23, Thm. 4.10, p. 205] for a solution of Exer. 4.7#1.

Exercise 4.7#7 is [24, Cor. A(2)].

Exercise 4.7#8 is [20, Prop. 3.2].

§4.8. This material is standard in the theory of "arithmetic groups." (If G is defined over $\mathbb{Q}$, then $G \cap SL(\ell, \mathbb{Z})$ is said to be an *arithmetic group*.) The book of Platonov and Rapinchuk [23] is an excellent reference on the subject. See [21] for an introduction. There are also numerous other books and survey papers.

Theorem 4.8.4 is due to A. Borel and Harish-Chandra [6]. (Many special cases had previously been treated by C. L. Siegel [26].) Expositions can also be found in [4, Cor. 13.2] and [23, Thm. 4.13]. (A proof of only Cor. 4.8.5 appears in [2, §V.2].) These are based on the reduction theory for arithmetic groups, not unipotent flows.

The observation that Thm. 4.8.4 can be obtained from a variation of Thm. 1.9.2 is due to G. A. Margulis [18, Rem. 3.12(II)].

§4.9. There are many textbooks on Lie groups, including [11, 12, 27]. The expository article of R. Howe [14] provides an elementary introduction.

References

[1] W. J. Baily, Jr.: *Introductory Lectures on Automorphic Forms.* Princeton University Press, Princeton, 1973. ISBN 0-691-08123-9, MR 51 #5982

[2] B. Bekka and M. Mayer: *Ergodic theory and Topological Dynamics of Group Actions on Homogeneous Spaces.* London Math. Soc. Lec. Notes #269. Cambridge U. Press, Cambridge, 2000. ISBN 0-521-66030-0, MR 2002c:37002

[3] A. Borel: Density properties for certain subgroups of semisimple groups without compact components. *Ann. of Math.* 72 (1960), 179–188. MR 23 #A964

[4] A. Borel: *Introduction aux Groupes Arithmétiques.* Actualités Scientifiques et Industrielles, No. 1341. Hermann, Paris, 1969. MR 39 #5577

[5] A. Borel: *Linear Algebraic Groups, 2nd ed.* Springer, New York, 1991. ISBN 0-387-97370-2, MR 92d:20001

[6] A. Borel and Harish-Chandra: Arithmetic subgroups of algebraic groups. *Ann. of Math.* 75 (1962), 485–535. MR 26 #5081

[7] A. Borel and G. Prasad: Values of isotropic quadratic forms at S-integral points. *Compositio Math.* 83 (1992), no. 3, 347–372. MR 93j:11022

[8] C. Chevalley: *Théorie des Groupes de Lie, Tome II.* Actualités Sci. Ind. no. 1152. Hermann, Paris, 1951. MR 14,448d

[9] C. Chevalley: *Théorie des Groupes de Lie, Tome III.* Actualités Sci. Ind. no. 1226. Hermann, Paris, 1955. MR 16,901a

[10] S. G. Dani: On ergodic quasi-invariant measures of group automorphism. *Israel J. Mathematics* 43 (1982) 62–74. MR 85d:22017

[11] S. Helgason: *Differential Geometry, Lie Groups, and Symmetric Spaces.* Pure and Applied Mathematics #80. Academic Press, New York, 1978. ISBN 0-12-338460-5, MR 80k:53081

[12] G. Hochschild: *The Structure of Lie Groups.* Holden-Day, San Francisco, 1965. MR 34 #7696

[13] G. P. Hochschild: *Basic Theory of Algebraic Groups and Lie Algebras.* Springer, New York, 1981. ISBN 0-387-90541-3, MR 82i:20002

[14] R. Howe: Very basic Lie theory. *Amer. Math. Monthly* 90 (1983), no. 9, 600–623; correction 91 (1984), no. 4, 247. MR 86m:00001a, 86m:00001b

[15] J. E. Humphreys: *Introduction to Lie Algebras and Representation Theory*. Springer, New York, 1972. ISBN 0-387-90053-5, MR 48 #2197

[16] J. E. Humphreys: *Linear Algebraic Groups*. Springer, New York, 1975. ISBN 0-387-90108-6, MR 53 #633

[17] S. Lang: *Algebra*, 2nd ed. Addison-Wesley, Reading, MA, 1984. ISBN 0-201-05487-6, MR 86j:00003

[18] G.A. Margulis: Lie groups and ergodic theory, in: L. Avramov and K. B. Tchakerian, eds., *Algebra—Some Current Trends* (*Varna, 1986*), pp. 130–146. Lecture Notes in Math. #1352, Springer, Berlin, 1988. ISBN 3-540-50371-4, MR 91a:22009

[19] G.A. Margulis: *Discrete Subgroups of Semisimple Lie Groups*. Springer, Berlin, 1991 ISBN 3-540-12179-X, MR 92h:22021

[20] G. A. Margulis and G. M. Tomanov: Invariant measures for actions of unipotent groups over local fields on homogeneous spaces. *Invent. Math.* 116 (1994) 347–392. MR 95k:22013

[21] D. Morris: *Introduction to Arithmetic Groups*. (preliminary version) http://arxiv.org/math/0106063

[22] A. L. Onishchik and E. B. Vinberg: *Lie Groups and Algebraic Groups*. Springer-Verlag, Berlin, 1990. ISBN 3-540-50614-4, MR 91g:22001

[23] V. Platonov and A. Rapinchuk: *Algebraic Groups and Number Theory*. Academic Press, Boston, 1994. ISBN 0-12-558180-7, MR 95b:11039

[24] M. Ratner: Raghunathan's topological conjecture and distributions of unipotent flows. *Duke Math. J.* 63 (1991), no. 1, 235–280. MR 93f:22012

[25] J.-P. Serre: *Complex Semisimple Lie Algebras*. Springer, Berlin, 2001. ISBN 3-540-67827-1, MR 2001h:17001

[26] C. L. Siegel: Einheiten quadratischer Formen. *Abh. Math. Sem. Hansischen Univ.* 13, (1940), 209–239. MR 2,148b

[27] V. S. Varadarajan: *Lie Groups, Lie Algebras, and Their Representations*. Springer, New York, 1984. ISBN 0-387-90969-9, MR 85e:22001

[28] D. Witte: Superrigidity of lattices in solvable Lie groups. *Invent. Math.* 122 (1995), no. 1, 147–193. MR 96k:22024

[29] R. J. Zimmer: *Ergodic Theory and Semisimple Groups*. Birkhäuser, Boston, 1984. ISBN 3-7643-3184-4, MR 86j:22014

CHAPTER 5

Proof of the Measure-Classification Theorem

In this chapter, we present the main ideas in a proof of the following theorem. The reader is assumed to be familiar with the concepts presented in Chap. 1.

(5.0.1) **Theorem** (Ratner). *If*
- *G is a closed, connected subgroup of $\mathrm{SL}(\ell, \mathbb{R})$, for some ℓ,*
- *Γ is a discrete subgroup of G,*
- *u^t is a unipotent one-parameter subgroup of G, and*
- *μ is an ergodic u^t-invariant probability measure on $\Gamma \backslash G$,*

*then μ is **homogeneous**.*

More precisely, there exist
- *a closed, connected subgroup S of G, and*
- *a point x in $\Gamma \backslash G$,*

such that

1) *μ is S-invariant, and*

2) *μ is supported on the orbit xS.*

(5.0.2) **Remark.** If we write $x = \Gamma g$, for some $g \in G$, and let $\Gamma_S = (g^{-1}\Gamma g) \cap S$, then the conclusions imply that

1) under the natural identification of the orbit xS with the homogeneous space $\Gamma_S \backslash S$, the measure μ is the Haar measure on $\Gamma_S \backslash S$,

2) Γ_S is a lattice in S, and

3) xS is closed

(see Exer. 1).

(5.0.3) **Assumption.** Later (see Assump. 5.3.1), in order to simplify the details of the proof while losing very few of the main ideas, we will make the additional assumption that

1) μ is invariant under a hyperbolic one-parameter subgroup $\{a^s\}$ that normalizes u^t, and

2) $\langle a^s, u^t \rangle$ is contained in a subgroup $L = \langle u^t, a^s, v^r \rangle$ that is locally isomorphic to SL(2, $\mathbb{R}$).

See §5.9 for a discussion of the changes involved in removing this hypothesis. The basic idea is that Prop. 1.6.10 shows that we may assume $\text{Stab}_G(\mu)$ contains a one-parameter subgroup that is not unipotent. A more sophisticated version of this argument, using the theory of algebraic groups, shows that slightly weakened forms of (1) and (2) are true. Making these assumptions from the start simplifies a lot of the algebra, without losing *any* of the significant ideas from dynamics.

(5.0.4) **Remark.** Note that G is *not* assumed to be semisimple. Although the semisimple case is the most interesting, we allow ourselves more freedom, principally because the proof relies (at one point, in the proof of Thm. 5.7.2) on induction on dim G, and this induction is based on knowing the result for all connected subgroups, not only the semisimple ones.

(5.0.5) **Remark.** There is no harm in assuming that G is almost Zariski closed (see Exer. 3). This provides a slight simplification in a couple of places (see Exer. 5.4#6 and the proof of Thm. 5.7.2).

Exercises for §5.0.

#1. Prove the assertions of Rem. 5.0.2 from the conclusions of Thm. 5.0.1.

#2. Show that Thm. 5.0.1 remains true without the assumption that G is connected.
[*Hint:* μ must be supported on a single connected component of $\Gamma \backslash G$. Apply Thm. 5.0.1 with $G°$ in the place of G.]

#3. Assume Thm. 5.0.1 is true under the additional hypothesis that G is almost Zariski closed. Prove that this additional hypothesis can be eliminated.
[*Hint:* $\Gamma \backslash G$ embeds in $\Gamma \backslash \text{SL}(\ell, \mathbb{R})$.]

5.1. An outline of the proof

Here are the main steps in the proof.

1) *Notation.*
- Let $S = \text{Stab}_G(\mu)$. We wish to show that μ is supported on a single S-orbit.
- Let $\mathfrak{g}$ be the Lie algebra of G and $\mathfrak{s}$ be the Lie algebra of S.

- The expanding and contracting subspaces of a^s (for $s > 0$) provide decompositions

$$\mathfrak{g} = \mathfrak{g}_- + \mathfrak{g}_0 + \mathfrak{g}_+ \quad \text{and} \quad \mathfrak{s} = \mathfrak{s}_- + \mathfrak{s}_0 + \mathfrak{s}_+,$$

and we have corresponding subgroups G_-, G_0, G_+, S_-, S_0, and S_+ (see Notn. 5.3.3).

- For convenience, let $U = S_+$. Note that U is unipotent, and we may assume $\{u^t\} \subset U$, so μ is ergodic for U.

2) We are interested in **transverse** divergence of nearby orbits. (We ignore relative motion *along* the U-orbits, and project to $G \ominus U$.) The shearing property of unipotent flows implies, for a.e. $x, y \in \Gamma \backslash G$, that if $x \approx y$, then the transverse divergence of the U-orbits through x and y is fastest along some direction in S (see Prop. 5.2.4). Therefore, the direction belongs to $G_- G_0$ (see Cor. 5.3.4).

3) We define a certain subgroup

$$\widetilde{S}_- = \{ g \in G_- \mid \forall u \in U, \ u^{-1} g u \in G_- G_0 U \}$$

of G_- (cf. Defn. 5.4.1). Note that $S_- \subset \widetilde{S}_-$.

The motivation for this definition is that if $y \in x\widetilde{S}_-$, then all of the transverse divergence belongs to $G_- G_0$ — there is no G_+-component to any of the transverse divergence. For clarity, we emphasize that this restriction applies to **all** transverse divergence, not only the **fastest** transverse divergence.

4) Combining (2) with the dilation provided by the translation a^{-s} shows, for a.e. $x, y \in \Gamma \backslash G$, that if $y \in xG_-$, then $y \in x\widetilde{S}_-$ (see Cor. 5.5.2).

5) A Lie algebra calculation shows that if $y \approx x$, and $y = xg$, with $g \in (G_- \ominus \widetilde{S}_-)G_0 G_+$, then the transverse divergence of the U-orbits through x and y is fastest along some direction in G_+ (see Lem. 5.5.3).

6) Because the conclusions of (2) and (5) are contradictory, we see, for a.e. $x, y \in \Gamma \backslash G$, that

$$\text{if } x \approx y, \text{ then } y \notin x(G_- \ominus \widetilde{S}_-)G_0 G_+$$

(cf. Cor. 5.5.4). (Actually, a technical problem causes us obtain this result only for x and y in a set of measure $1 - \epsilon$.)

7) The relation between stretching and entropy (Prop. 2.5.11) provides bounds on the entropy of a^s, in terms of the the

Jacobian of a^s on U and (using (4)) the Jacobian of a^{-s} on $\widetilde{S}_-$:

$$J(a^s, U) \le h_\mu(a^s) \le J(a^{-s}, \widetilde{S}_-).$$

On the other hand, the structure of $\mathfrak{sl}(2, \mathbb{R})$-modules implies that $J(a^s, U) \ge J(a^{-s}, \widetilde{S}_-)$. Thus, we conclude that $h_\mu(a^s) = J(a^{-s}, \widetilde{S}_-)$. This implies that $\widetilde{S}_- \subset \mathrm{Stab}_G(\mu)$, so we must have $\widetilde{S}_- = S_-$ (see Prop. 5.6.1).

8) By combining the conclusions of (6) and (7), we show that $\mu(xS_-G_0G_+) > 0$, for some $x \in \Gamma\backslash G$ (see Prop. 5.7.1).

9) By combining (8) with the (harmless) assumption that μ is not supported on an orbit of any closed, proper subgroup of G, we show that $S_- = G_-$ (so S_- is horospherical), and then there are a number of ways to show that $S = G$ (see Thm. 5.7.2).

The following several sections expand this outline into a fairly complete proof, modulo some details that are postponed to §5.8.

5.2. Shearing and polynomial divergence

As we saw in Chap. 1, shearing and polynomial divergence are crucial ingredients of the proof of Thm. 5.0.1. Precise statements will be given in §5.8, but let us now describe them informally. Our goal here is to prove that the direction of fastest divergence usually belongs to the stabilizer of μ (see Prop. 5.2.4′, which follows Cor. 5.2.5). This will later be restated in a slightly more convenient (but weaker) form (see Cor. 5.3.4).

(5.2.1) **Lemma** (Shearing). *If U is any connected, unipotent subgroup of G, then the transverse divergence of any two nearby U-orbits is fastest along some direction that is in the normalizer $N_G(U)$.*

(5.2.2) **Lemma** (Polynomial divergence). *If U is a connected, unipotent subgroup of G, then any two nearby U-orbits diverge at polynomial speed.*

Hence, if it takes a certain amount of time for two nearby U-orbits to diverge to a certain distance, then the amount (and direction) of divergence will remain approximately the same for a proportional length of time.

By combining these two results we will establish the following conclusion (cf. Cor. 1.6.9). It is the basis of the entire proof.

(5.2.3) **Notation.** Let $S = \mathrm{Stab}_G(\mu)^\circ$. This is a closed subgroup of G (see Exer. 1).

(5.2.4) **Proposition.** *If U is any connected, ergodic, unipotent subgroup of S, then there is a conull subset Ω of $\Gamma \backslash G$, such that, for all $x, y \in \Omega$, with $x \approx y$, the U-orbits through x and y diverge fastest along some direction that belongs to S.*

This immediately implies the following interesting special case of Ratner's Theorem (see Exer. 4), which was proved rather informally in Chap. 1 (see Prop. 1.6.10).

(5.2.5) **Corollary.** *If $U = \mathrm{Stab}_G(\mu)$ is unipotent (and connected), then μ is supported on a single U-orbit.*

Although Prop. 5.2.4 is true (see Exer. 5), it seems to be very difficult to prove from scratch, so we will be content with proving the following weaker version that does not yield a conull subset, and imposes a restriction on the relation between x and y (see 5.8.8). (See Exer. 5.8#5 for a non-infinitesimal version of the result.)

(5.2.4′) **Proposition.** *For any*

- *connected, ergodic, unipotent subgroup U of S, and*
- *any $\epsilon > 0$,*

there is a subset Ω_ϵ of $\Gamma \backslash G$, such that

1) *$\mu(\Omega_\epsilon) > 1 - \epsilon$, and*

2) *for all $x, y \in \Omega_\epsilon$, with $x \approx y$, and such that a certain technical assumption (5.8.9) is satisfied, the fastest transverse divergence of the U-orbits through x and y is along some direction that belongs to S.*

Proof (cf. Cor. 1.6.5). Let us assume that no $N_G(U)$-orbit has positive measure, for otherwise it is easy to complete the proof (cf. Exer. 3). Then, for a.e. $x \in \Gamma \backslash G$, there is a point $y \approx x$, such that

1) $y \notin x\, N_G(U)$, and

2) y is a generic point for μ (see Cor. 3.1.6).

Because $y \notin x\, N_G(U)$, we know that the orbit yU is not parallel to xU, so they diverge from each other. From Lem. 5.2.1, we know that the direction of fastest transverse divergence belongs to $N_G(U)$, so there exist $u, u' \in U$, and $c \in N_G(U) \ominus U$, such that

- $yu' \approx (xu)c$, and
- $\|c\| \asymp 1$ (i.e., $\|c\|$ is finite, but not infinitesimal).

Because $c \notin U = \mathrm{Stab}_G(\mu)$, we know that $c_*\mu \neq \mu$. Because $c \in N_G(U)$, this implies $c_*\mu \perp \mu$ (see Exer. 6), so there is a compact subset K with $\mu(K) > 1 - \epsilon$ and $K \cap Kc = \varnothing$ (see Exer. 7).

We would like to complete the proof by saying that there are values of u for which both of the two points xu and yu' are arbitrarily close to K, which contradicts the fact that $d(K, Kc) > 0$. However, there are two technical problems:

1) The set K must be chosen before we know the value of c. This issue is handled by Lem. 5.8.6.

2) The Pointwise Ergodic Theorem (3.4.3) implies (for a.e. x) that xu is arbitrarily close to K a huge proportion of the time. But this theorem does not apply directly to yu', because u' is a nontrivial function of u. To overcome this difficulty, we add an additional technical hypothesis on the element g with $y = xg$ (see 5.8.8). With this assumption, the result can be proved (see 5.8.7), by showing that the Jacobian of the change of variables $u \mapsto u'$ is bounded above and below on some set of reasonable size, and applying the uniform approximate version of the Pointwise Ergodic Theorem (see Cor. 3.4.4). The uniform estimate is what requires us to restrict to a set of measure $1 - \epsilon$, rather than a conull set.   □

(5.2.6) Remark.

1) The fact that Ω_ϵ is not quite conull is not a serious problem, although it does make one part of the proof more complicated (cf. Prop. 5.7.1).

2) We will apply Prop. 5.2.4′ only twice (in the proofs of Cors. 5.5.2 and 5.5.4). In each case, it is not difficult to verify that the technical assumption is satisfied (see Exers. 5.8#1 and 5.8#2).

Exercises for §5.2.

#1. Show that $\mathrm{Stab}_G(\mu)$ is a closed subgroup of G.
[*Hint:* $g \in \mathrm{Stab}_G(\mu)$ if and only if $\int f(xg)\,d\mu(x) = \int f\,d\mu$ for all continuous functions f with compact support.]

#2. Suppose
- ν is a (finite or infinite) Borel measure on G, and
- N is a unimodular, normal subgroup of G.

Show that if ν is right-invariant under N (that is, $\nu(An) = \nu(A)$ for all $n \in N$), then ν is left-invariant under N.

#3. Show that if
- N is a unimodular, normal subgroup of G,

- N is contained in $\mathrm{Stab}_G(\mu)$, and
- N is ergodic on $\Gamma \backslash G$,

then μ is homogeneous.

[*Hint:* Lift μ to an (infinite) measure $\hat{\mu}$ on G, such that $\hat{\mu}$ is left invariant under Γ, and right invariant under N. Exercise 2 implies that $\hat{\mu}$ is left invariant (and ergodic) under the closure H of ΓN. Ergodicity implies that $\hat{\mu}$ is supported on a single H-orbit.]

#4. Prove Cor. 5.2.5 from Prop. 5.2.4 and Exer. 3.

[*Hint:* If $\mu(xN_G(U)) > 0$, for some $x \in \Gamma \backslash G$, then Exer. 3 (with $N_G(U)$ in the place of G) implies that μ is homogeneous. Otherwise, Prop. 5.2.4 implies that $\mathrm{Stab}_G(\mu) \smallsetminus U \neq \varnothing$.]

#5. Show that Thm. 5.0.1 implies Prop. 5.2.4.

#6. Show that if

- μ is U-invariant and ergodic, and
- $c \in N_G(U)$,

then

(a) $c_* \mu$ is U-invariant and ergodic, and

(b) either $c_* \mu = \mu$ or $c_* \mu \perp \mu$.

#7. Suppose

- $\epsilon > 0$,
- μ is U-invariant and ergodic,
- $c \in N_G(U)$, and
- $c_* \mu \perp \mu$.

Show that there is a compact subset K of $\Gamma \backslash G$, such that

(a) $\mu(K) > 1 - \epsilon$, and

(b) $K \cap Kc = \varnothing$.

5.3. Assumptions and a restatement of 5.2.4′

(5.3.1) **Assumption.** As mentioned in Assump. 5.0.3, we assume there exist

- a closed subgroup L of G and
- a (nontrivial) one-parameter subgroup $\{a^s\}$ of L,

such that

1) $\{u^t\} \subset L$,

2) $\{a^s\}$ is hyperbolic, and normalizes $\{u^t\}$,

3) μ is invariant under $\{a^s\}$, and

4) L is locally isomorphic to $\mathrm{SL}(2, \mathbb{R})$.

(5.3.2) **Remark.**

1) Under an appropriate local isomorphism between L and $\mathrm{SL}(2,\mathbb{R})$, the subgroup $\langle a^s, u^t \rangle$ maps to the group $\mathbb{D}_2 \mathbb{U}_2$ of lower triangular matrices in $\mathrm{SL}(2,\mathbb{R})$ (see Exer. 4.9#1).

2) Therefore, the parametrizations of a^s and u^t can be chosen so that $a^{-s} u^t a^s = u^{e^{2s}t}$ for all s and t.

3) The Mautner Phenomenon implies that the measure μ is ergodic for $\{a^s\}$ (see Cor. 3.2.5).

(5.3.3) **Notation.**

- For a (small) element g of G, we use $\underline{g}$ to denote the corresponding element $\log g$ of the Lie algebra $\mathfrak{g}$.

- Recall that $S = \mathrm{Stab}_G(\mu)^\circ$ (see Notn. 5.2.3).

- By renormalizing, let us assume that $[\underline{u}, \underline{a}] = 2\underline{u}$ (where $a = a^1$ and $u = u^1$).

- Let $\{v^r\}$ be the (unique) one-parameter unipotent subgroup of L, such that $[\underline{v}, \underline{a}] = -2\underline{v}$ and $[\underline{v}, \underline{u}] = \underline{a}$ (see Eg. 4.9.11).

- Let $\bigoplus_{\lambda \in \mathbb{Z}} \mathfrak{g}_\lambda$ be the decomposition of $\mathfrak{g}$ into weight spaces of $\underline{a}$: that is,

$$\mathfrak{g}_\lambda = \left\{ \underline{g} \in \mathfrak{g} \mid [\underline{g}, \underline{a}] = \lambda \underline{g} \right\}.$$

- Let $\mathfrak{g}_+ = \bigoplus_{\lambda > 0} \mathfrak{g}_\lambda$, $\mathfrak{g}_- = \bigoplus_{\lambda < 0} \mathfrak{g}_\lambda$, $\mathfrak{s}_+ = \mathfrak{s} \cap \mathfrak{g}_+$, $\mathfrak{s}_- = \mathfrak{s} \cap \mathfrak{g}_-$, and $\mathfrak{s}_0 = \mathfrak{s} \cap \mathfrak{g}_0$. Then

$$\mathfrak{g} = \mathfrak{g}_- + \mathfrak{g}_0 + \mathfrak{g}_+ \qquad \text{and} \qquad \mathfrak{s} = \mathfrak{s}_- + \mathfrak{s}_0 + \mathfrak{s}_+.$$

These are direct sums of vector spaces, although they are not direct sums of Lie algebras.

- Let $G_+, G_-, G_0, S_+, S_-, S_0$ be the connected subgroups of G corresponding to the Lie subalgebras $\mathfrak{g}_+, \mathfrak{g}_-, \mathfrak{g}_0, \mathfrak{s}_+, \mathfrak{s}_-, \mathfrak{s}_0$, respectively (see Exer. 1).

- Let $U = S_+$ (and let $\mathfrak{u}$ be the Lie algebra of U).

Because $S_- S_0 U = S_- S_0 S_+$ contains a neighborhood of e in S (see Exer. 2), Prop. 5.2.4′ states that the direction of fastest transverse divergence belongs to $S_- S_0$. The following corollary is *a priori* weaker (because G_- and G_0 are presumably larger than S_- and S_0), but it is the only consequence of Lem. 5.8.6 or Lem. 5.2.1 that we will need in our later arguments.

(5.3.4) **Corollary.** *For any $\epsilon > 0$, there is a subset Ω_ϵ of $\Gamma \backslash G$, such that*

1) $\mu(\Omega_\epsilon) > 1 - \epsilon$, *and*

2) *for all $x, y \in \Omega_\epsilon$, with $x \approx y$, and such that a certain technical assumption (5.8.9) is satisfied, the fastest transverse divergence of the U-orbits through x and y is along some direction that belongs to $G_- G_0$.*

Exercises for §5.3.

#1. Show $\mathfrak{g}_+$, $\mathfrak{g}_-$, and $\mathfrak{g}_0$ are subalgebras of $\mathfrak{g}$.
[*Hint:* $[\mathfrak{g}_{\lambda_1}, \mathfrak{g}_{\lambda_2}] \subset \mathfrak{g}_{\lambda_1 + \lambda_2}$.]

#2. Show $S_- S_0 S_+$ contains a neighborhood of e in S.
[*Hint:* Because $\mathfrak{s}_- + \mathfrak{s}_0 + \mathfrak{s}_+ = \mathfrak{s}$, this follows from the Inverse Function Theorem.]

5.4. Definition of the subgroup $\widetilde{S}$

To exploit Cor. 5.3.4, let us introduce some notation. The corollary states that orbits diverge **fastest** along some direction in $G_- G_0$, but it will be important to understand when **all** of the transverse divergence, not just the fastest part, is along $G_- G_0$. More precisely, we wish to understand the elements g of G, such that if $y = xg$, then the orbits through x and y diverge transversely only along directions in $G_- G_0$: the G_+-component of the relative motion should belong to U, so the G_+-component of the divergence is trivial. Because the divergence is measured by $u^{-1} g u$ (thought of as an element of G/U), this suggests that we wish to understand

$$\left\{ g \in G \; \middle| \; \begin{array}{c} u^{-1} g u \in G_- G_0 U, \\ \forall u \in \text{some neighborhood of } e \text{ in } U \end{array} \right\}.$$

This is the right idea, but replacing $G_- G_0 U$ with its Zariski closure $\overline{G_- G_0 U}$ yields a slightly better theory. (For example, the resulting subset of G turns out to be a subgroup!) Fortunately, when g is close to e (which is the case we are usually interested in), this alteration of the definition makes no difference at all (see Exer. 10). (This is because $G_- G_0 U$ contains a neighborhood of e in $\overline{G_- G_0 U}$ (see Exer. 6).) Thus, the non-expert may wish to think of $\overline{G_- G_0 U}$ as simply being $G_- G_0 U$, although this is not strictly correct.

(5.4.1) **Definition.** Let

$$\widetilde{S} = \left\{ g \in G \; \middle| \; u^{-1} g u \in \overline{G_- G_0 U}, \text{ for all } u \in U \right\}$$

and

$$\widetilde{S}_- = \widetilde{S} \cap G_-.$$

It is more or less obvious that $S \subset \tilde{S}$ (see Exer. 4). Although this is much less obvious, it should also be noted that $\tilde{S}$ is a closed subgroup of G (see Exer. 8).

(5.4.2) **Remark.** Here is an alternate approach to the definition of $\tilde{S}$, or, at least, its identity component.

1) Let

$$\tilde{\mathfrak{s}} = \left\{ \underline{g} \in \mathfrak{g} \mid \underline{g}(\operatorname{ad}\underline{u})^k \in \mathfrak{g}_- + \mathfrak{g}_0 + \mathfrak{u}, \ \forall k \geq 0, \ \forall \underline{u} \in \mathfrak{u} \right\}.$$

Then $\tilde{\mathfrak{s}}$ is a Lie subalgebra of $\mathfrak{g}$ (see Exer. 11), so we may let $\tilde{S}^\circ$ be the corresponding connected Lie subgroup of G. (We will see in (3) below that this agrees with Defn. 5.4.1.)

2) From the point of view in (1), it is not difficult to see that $\tilde{S}^\circ$ is the unique maximal connected subgroup of G, such that

(a) $\tilde{S}^\circ \cap G_+ = U$, and

(b) $\tilde{S}^\circ$ is normalized by a^t

(see Exers. 12 and 13). This makes it obvious that $S \subset \tilde{S}^\circ$. It is also easy to verify directly that $\mathfrak{s} \subset \tilde{\mathfrak{s}}$ (see Exer. 14).

3) It is not difficult to see that the identity component of the subgroup defined in Defn. 5.4.1 is also the subgroup characterized in (2) (see Exer. 15), so this alternate approach agrees with the original definition of $\tilde{S}$.

(5.4.3) **Example.** Remark 5.4.2 makes it easy to calculate $\tilde{S}^\circ$.

1) We have $\tilde{S} = G$ if and only if $U = G_+$ (see Exer. 16).

2) If

$$G = \operatorname{SL}(3, \mathbb{R}), \quad \underline{a} = \begin{bmatrix} 1 & 0 & 0 \\ 0 & 0 & 0 \\ 0 & 0 & -1 \end{bmatrix}, \quad \text{and} \quad \mathfrak{u} = \begin{bmatrix} 0 & 0 & 0 \\ 0 & 0 & 0 \\ * & 0 & 0 \end{bmatrix},$$

then

$$\mathfrak{g}_+ = \begin{bmatrix} 0 & 0 & 0 \\ * & 0 & 0 \\ * & * & 0 \end{bmatrix} \quad \text{and} \quad \tilde{\mathfrak{s}} = \begin{bmatrix} * & 0 & * \\ 0 & * & 0 \\ * & 0 & * \end{bmatrix}$$

(see Exer. 17).

3) If

$$G = \operatorname{SL}(3, \mathbb{R}), \quad \underline{a} = \begin{bmatrix} 1 & 0 & 0 \\ 0 & 0 & 0 \\ 0 & 0 & -1 \end{bmatrix}, \quad \text{and} \quad \mathfrak{u} = \begin{bmatrix} 0 & 0 & 0 \\ * & 0 & 0 \\ * & 0 & 0 \end{bmatrix},$$

then

$$\mathfrak{g}_+ = \begin{bmatrix} 0 & 0 & 0 \\ * & 0 & 0 \\ * & * & 0 \end{bmatrix} \quad \text{and} \quad \widetilde{\mathfrak{s}} = \begin{bmatrix} * & 0 & * \\ * & * & * \\ * & 0 & * \end{bmatrix}$$

(see Exer. 18).

4) If

$$G = \mathrm{SL}(3, \mathbb{R}), \quad \underline{a} = \begin{bmatrix} 2 & 0 & 0 \\ 0 & 0 & 0 \\ 0 & 0 & -2 \end{bmatrix}, \quad \text{and} \quad \mathfrak{u} = \mathbb{R} \begin{bmatrix} 0 & 0 & 0 \\ 1 & 0 & 0 \\ 0 & 1 & 0 \end{bmatrix},$$

then

$$\mathfrak{g}_+ = \begin{bmatrix} 0 & 0 & 0 \\ * & 0 & 0 \\ * & * & 0 \end{bmatrix} \quad \text{and} \quad \widetilde{\mathfrak{s}} = \mathbb{R} \begin{bmatrix} 0 & 1 & 0 \\ 0 & 0 & 1 \\ 0 & 0 & 0 \end{bmatrix} + \begin{bmatrix} * & 0 & 0 \\ 0 & * & 0 \\ 0 & 0 & * \end{bmatrix} + \mathfrak{u}$$

(see Exer. 19).

5) If

$$G = \mathrm{SL}(2, \mathbb{R}) \times \mathrm{SL}(2, \mathbb{R}), \quad \underline{a} = \left(\begin{bmatrix} 1 & 0 \\ 0 & -1 \end{bmatrix}, \begin{bmatrix} 1 & 0 \\ 0 & -1 \end{bmatrix} \right),$$

and

$$\mathfrak{u} = \mathbb{R} \left(\begin{bmatrix} 0 & 0 \\ 1 & 0 \end{bmatrix}, \begin{bmatrix} 0 & 0 \\ 1 & 0 \end{bmatrix} \right),$$

then

$$\mathfrak{g}_+ = \begin{bmatrix} 0 & 0 \\ * & 0 \end{bmatrix} \times \begin{bmatrix} 0 & 0 \\ * & 0 \end{bmatrix} \quad \text{and} \quad \widetilde{\mathfrak{s}} = \mathbb{R} \left(\begin{bmatrix} 0 & 1 \\ 0 & 0 \end{bmatrix}, \begin{bmatrix} 0 & 1 \\ 0 & 0 \end{bmatrix} \right) + \mathbb{R}\underline{a} + \mathfrak{u}$$

(see Exer. 20).

Exercises for §5.4.

#1. Show that if
- V is any subgroup of G_+ (or of G_-), and
- V is normalized by $\{a^t\}$,

then V is connected.
[*Hint:* If $v \in G_+$, then $a^{-t} v a^t \to e$ as $t \to -\infty$.]

#2. Show that if H is a connected subgroup of G, and H is normalized by $\{a^t\}$, then $H \subset \overline{\overline{H_- H_0 H_+}}$.
[*Hint:* $\dim \overline{\overline{H_-}} \, \overline{\overline{H_0}} \, \overline{\overline{H_+}} = \dim \overline{\overline{H}}$. Use Exer. 4.1#11.]

#3. Show, directly from Defn. 5.4.1, that $N_{G_-}(U) \subset \widetilde{S}_-$.

#4. Show, directly from Defn. 5.4.1, that $S \subset \widetilde{S}$.
[*Hint:* Use Exer. 2.]

#5. Let $G = \mathrm{SL}(2, \mathbb{R})$, $a^t = \begin{bmatrix} e^t & 0 \\ 0 & e^{-t} \end{bmatrix}$ and $U = G_+ = \begin{bmatrix} 1 & 0 \\ * & 1 \end{bmatrix}$.

(a) Show that $G_- G_0 G_+ \neq G$.

(b) For $g \in G$, show that if $u^{-1} g u \in G_- G_0 U$, for all $u \in U$, then $g \in G_0 U$.

(c) Show, for all $g \in G$, and all $u \in U$, that $u^{-1} g u \in \overline{\overline{G_- G_0 U}}$. Therefore $\tilde{S} = G$.

[*Hint:* Letting $v = (0,1)$, and considering the usual representation of G on $\mathbb{R}^2$, we have $U = \mathrm{Stab}_G(v)$. Thus, G/U may be identified with $\mathbb{R}^2 \smallsetminus \{0\}$. This identifies $G_- G_0 U / U$ with $\{(x,y) \in \mathbb{R}^2 \mid x > 0\}$.]

#6. Assume G is almost Zariski closed (see Rem. 5.0.5). Define the polynomial $\psi \colon G_- G_0 \times G_+ \to G_- G_0 G_+$ by $\psi(g, u) = gu$. (Note that $G_- G_0 G_+$ is an open subset of G (cf. Exer. 5.3#2).) Assume the inverse of ψ is rational (although we do not prove it, this is indeed always the case, cf. Exer. 5).

Show that $G_- G_0 U$ is an open subset of $\overline{\overline{G_- G_0 U}}$.

[*Hint:* $G_- G_0 U$ is the inverse image of U under a rational map $\psi_+^{-1} \colon G_- G_0 G_+ \to G_+$.]

#7. (a) Show that if
- V is a Zariski closed subset of $\mathrm{SL}(\ell, \mathbb{R})$,
- $g \in \mathrm{SL}(\ell, \mathbb{R})$, and
- $Vg \subset V$,

then $Vg = V$.

(b) Show that if V is a Zariski closed subset of $\mathrm{SL}(\ell, \mathbb{R})$, then
$$\{ g \in \mathrm{SL}(\ell, \mathbb{R}) \mid Vg \subset V \}$$
is a closed subgroup of $\mathrm{SL}(\ell, \mathbb{R})$.

(c) Construct an example to show that the conclusion of (7b) can fail if V is assumed only to be closed, not Zariski closed.
[*Hint:* Use Exer. 4.1#11.]

#8. Show, directly from Defn. 5.4.1, that $\tilde{S}$ is a subgroup of G.
[*Hint:* Show that
$$\tilde{S} = \left\{ g \in G \;\middle|\; \overline{\overline{G_- G_0 U}}\, g \subset \overline{\overline{G_- G_0 U}} \right\},$$
and apply Exer. 7b.]

#9. Show that if $g \notin \tilde{S}$, then
$$\{ u \in U \mid u^{-1} g u \in \overline{\overline{G_- G_0 U}} \}$$

is nowhere dense in U. That is, its closure does not contain any open subset of U.

[*Hint:* It is a Zariski closed, proper subset of U.]

#10. Show that there is a neighborhood W of e in G, such that

$$\tilde{S} \cap W = \left\{ g \in W \ \middle| \ \begin{array}{c} u^{-1}gu \in G_- G_0 U, \\ \forall u \in \text{ some neighborhood of } e \text{ in } U \end{array} \right\}.$$

[*Hint:* Use Exers. 9 and 6.]

#11. Show, directly from the definition (see 5.4.1), that

   (a) $\tilde{\mathfrak{s}}$ is invariant under ad $\underline{a}$, and

   (b) $\tilde{\mathfrak{s}}$ is a Lie subalgebra of $\mathfrak{g}$.

[*Hint:* If $\underline{g}_1 \in \tilde{\mathfrak{s}}_{\lambda_1}, \underline{g}_2 \in \tilde{\mathfrak{s}}_{\lambda_2}, \underline{u} \in \mathfrak{u}_{\lambda_3}$, and $\lambda_1 + \lambda_2 + (k_1 + k_2)\lambda_3 > 0$, then $\underline{g}_i (\mathrm{ad}\,\underline{u})^{k_i} \in \mathfrak{u}$, for some $i \in \{1, 2\}$, so

$$[\underline{g}_1 (\mathrm{ad}\,\underline{u})^{k_1}, \underline{g}_2 (\mathrm{ad}\,\underline{u})^{k_2}] \in \mathfrak{g}_- + \mathfrak{g}_0 + \mathfrak{u},$$

and it follows that $\tilde{\mathfrak{s}}$ is a Lie subalgebra.]

#12. Show, directly from Defn. 5.4.1, that

   (a) $\tilde{S} \cap G_+ = U$, and

   (b) $\tilde{S}$ is normalized by $\{a^t\}$.

[*Hint:* It suffices to show that $\tilde{\mathfrak{s}}_+ = \mathfrak{u}$ (see Exer. 1), and that $\tilde{\mathfrak{s}}$ is $(\mathrm{Ad}_G\, a^t)$-invariant.]

#13. Show, directly from Defn. 5.4.1, that if H is any connected subgroup of G, such that

   (a) $H \cap G_+ = U$, and

   (b) H is normalized by $\{a^t\}$,

then $H \subset \tilde{S}$.

[*Hint:* It suffices to show that $\mathfrak{h} \subset \tilde{\mathfrak{s}}$.]

#14. Show, directly from the definition of $\tilde{\mathfrak{s}}$ in Rem. 5.4.2(1), that $\mathfrak{s} \subset \tilde{\mathfrak{s}}$.

#15. Verify, directly from Defn. 5.4.1 (and assuming that $\tilde{S}$ is a subgroup),

   (a) that $\tilde{S}$ satisfies conditions (a) and (b) of Rem. 5.4.2(2), and

   (b) conversely, that if H is a connected subgroup of G, such that $H \cap G_+ = U$ and H is normalized by $\{a^t\}$, then $H \subset \tilde{S}$.

#16. Verify Eg. 5.4.3(1).

#17. Verify Eg. 5.4.3(2).

#18. Verify Eg. 5.4.3(3).

#19. Verify Eg. 5.4.3(4).

#20. Verify Eg. 5.4.3(5).

5.5. Two important consequences of shearing

Our ultimate goal is to find a conull subset Ω of $\Gamma\backslash G$, such that if $x, y \in \Omega$, then $y \in xS$. In this section, we establish two consequences of Cor. 5.3.4 that represent major progress toward this goal (see Cors. 5.5.2 and 5.5.4). These results deal with $\tilde{S}$, rather than S, but that turns out not to be a very serious problem, because $\tilde{S} \cap G_+ = S \cap G_+$ (see Rem. 5.4.2(2)) and $\tilde{S} \cap G_- = S \cap G_-$ (see Prop. 5.6.1).

(5.5.1) Notation. Let

- $\mathfrak{g}_+ \ominus \mathfrak{u}$ be an a^s-invariant complement to $\mathfrak{u}$ in $\mathfrak{g}_+$,
- $\mathfrak{g}_- \ominus \tilde{\mathfrak{s}}_-$ be an a^s-invariant complement to $\tilde{\mathfrak{s}}_-$ in $\mathfrak{g}_-$,
- $G_+ \ominus U = \exp(\mathfrak{g}_- \ominus \mathfrak{u})$,
  and
- $G_- \ominus \tilde{S}_- = \exp(\mathfrak{g}_- \ominus \tilde{\mathfrak{s}}_-)$.

Note that the natural maps $(G_+ \ominus U) \times U \to G_+$ and $(G_- \ominus \tilde{S}_-) \times \tilde{S}_- \to G_-$ (defined by $(g, h) \mapsto gh$) are diffeomorphisms (see Exer. 1).

(5.5.2) Corollary. *There is a conull subset Ω of $\Gamma\backslash G$, such that if $x, y \in \Omega$, and $y \in xG_-$, then $y \in x\tilde{S}_-$.*

Proof. Choose Ω_0 as in the conclusion of Cor. 5.3.4. From the Pointwise Ergodic Theorem (3.1.3), we know that

$$\Omega = \left\{ x \in \Gamma\backslash G \mid \{ t \in \mathbb{R}^+ \mid xa^t \in \Omega_0 \} \text{ is unbounded} \right\}$$

is conull (see Exer. 3.1#3).

We have $y = xg$, for some $g \in G_-$. Because $a^{-t}ga^t \to e$ as $t \to \infty$, we may assume, by replacing x and y with xa^t and ya^t for some infinitely large t, that g is infinitesimal (and that $x, y \in \Omega_0$). (See Exer. 5.8#6 for a non-infinitesimal version of the proof.)

Suppose $g \notin \tilde{S}_-$ (this will lead to a contradiction). From the definition of $\tilde{S}_-$, this means there is some $u \in U$, such that $u^{-1}gu \notin G_-G_0U$: write $u^{-1}gu = hcu'$ with $h \in G_-G_0$, $c \in G_+ \ominus U$, and $u' \in U$. We may assume h is infinitesimal (because we could choose u to be finite, or even infinitesimal, if desired (see Exer. 5.4#9)). Translating again by an (infinitely large) element of $\{a^t\}$, with $t \geq 0$, we may assume c is infinitely large. Because h is infinitesimal, this clearly implies that the orbits through x and y diverge fastest along a direction in G_+, not a direction in G_-G_0. This contradicts Cor. 5.3.4. (See Exer. 5.8#1 for a verification of the technical assumption (5.8.9) in that corollary.) $\qquad\square$

An easy calculation (involving only algebra, not dynamics) establishes the following. (See Exer. 5.8#7 for a non-infinitesimal version.)

(5.5.3) Lemma. *If*

- *$y = xg$ with*
- *$g \in (G_- \ominus \widetilde{S}_-)G_0G_+$, and*
- *$g \approx e$,*

then the transverse divergence of the U-orbits through x and y is fastest along some direction in G_+.

Proof. Choose $s > 0$ (infinitely large), such that $\hat{g} = a^s g a^{-s}$ is finite, but not infinitesimal, and write $\hat{g} = \hat{g}_-\hat{g}_0\hat{g}_+$, with $\hat{g}_- \in G_-, \hat{g}_0 \in G_0$, and $\hat{g}_+ \in G_+$. (Note that $\hat{g}_0$ and $\hat{g}_+$ are infinitesimal, but $\hat{g}_-$ is not.) Because $\hat{g}_- \in G_- \ominus \widetilde{S}_-$, we know that $\hat{g}$ is not infinitely close to $\widetilde{S}_-$, so there is some finite $u \in U$, such that $u^{-1}\hat{g}$ is **not** infinitesimally close to G_-G_0U.

Let $\hat{u} = a^{-s}ua^s$, and consider $\hat{u}^{-1}g\hat{u} = a^{-s}(u^{-1}\hat{g}u)a^s$.

- Because $u^{-1}\hat{g}u$ is finite (since u and $\hat{g}$ are finite), we know that each of $(u^{-1}\hat{g}u)_-$ and $(u^{-1}\hat{g}u)_0$ is finite. Therefore $(\hat{u}^{-1}g\hat{u})_-$ and $(\hat{u}^{-1}g\hat{u})_0$ are finite, because conjugation by a^s does not expand G_- or G_0.
- On the other hand, we know that $(\hat{u}^{-1}g\hat{u})_+$ is infinitely far from U, because the distance between $u^{-1}\hat{g}u$ and U is not infinitesimal, and conjugation by a^s expands G_+ by an infinite factor.

Therefore, the fastest divergence is clearly along a direction in G_+. $\qquad\square$

The conclusion of the above lemma contradicts the conclusion of Cor. 5.3.4(2) (and the technical assumption (5.8.9) is automatically satisfied in this situation (see Exer. 5.8#2)), so we have the following conclusion:

(5.5.4) Corollary. *For any $\epsilon > 0$, there is a subset Ω_ϵ of $\Gamma\backslash G$, such that*

1) *$\mu(\Omega_\epsilon) > 1 - \epsilon$, and*
2) *for all $x, y \in \Omega_\epsilon$, with $x \approx y$, we have $y \notin x(G_- \ominus \widetilde{S}_-)G_0G_+$.*

This can be restated in the following non-infinitesimal terms (see Exer. 5.8#8):

(5.5.4′) **Corollary.** *For any $\epsilon > 0$, there is a subset Ω_ϵ of $\Gamma \backslash G$, and some $\delta > 0$, such that*

1) $\mu(\Omega_\epsilon) > 1 - \epsilon$, *and*

2) *for all $x, y \in \Omega_\epsilon$, with $d(x, y) < \delta$, we have $y \notin x(G_- \ominus \widetilde{S}_-)G_0 G_+$.*

Exercise for §5.5.

#1. Show that if $\mathfrak{v}$ and $\mathfrak{w}$ are two complementary a^t-invariant subspaces of $\mathfrak{g}_+$, then the natural map $\exp \mathfrak{v} \times \exp \mathfrak{w} \to G_+$, defined by $(v, w) \mapsto vw$, is a diffeomorphism.
   [*Hint:* The Inverse Function Theorem implies that the map is a local diffeomorphism near e. Conjugate by a^s to expand the good neighborhood.]

5.6. Comparing $\widetilde{S}_-$ with S_-

We will now show that $\widetilde{S}_- = S_-$ (see Prop. 5.6.1). To do this, we use the following lemma on the entropy of translations on homogeneous spaces. Corollary 5.5.2 is what makes this possible, by verifying the hypotheses of Lem. 2.5.11′(2), with $W = \widetilde{S}_-$.

(2.5.11′) **Lemma.** *Suppose W is a closed, connected subgroup of G_- that is normalized by a, and let*

$$J(a^{-1}, W) = \det((\operatorname{Ad} a^{-1})|_\mathfrak{w})$$

be the Jacobian of a^{-1} on W.

1) *If μ is W-invariant, then $h_\mu(a) \geq \log J(a^{-1}, W)$.*

2) *If there is a conull, Borel subset Ω of $\Gamma \backslash G$, such that $\Omega \cap xG_- \subset xW$, for every $x \in \Omega$, then $h_\mu(a) \leq \log J(a^{-1}, W)$.*

3) *If the hypotheses of (2) are satisfied, and equality holds in its conclusion, then μ is W-invariant.*

(5.6.1) **Proposition.** *We have $\widetilde{S}_- = S_-$.*

Proof (cf. proofs of Cors. 1.7.6 and 1.8.1). We already know that $\widetilde{S}_- \supset S_-$ (see Rem. 5.4.2(2)). Thus, because $\widetilde{S}_- \subset G_-$, it suffices to show that $\widetilde{S}_- \subset S$. That is, it suffices to show that μ is $\widetilde{S}_-$-invariant.

From Lem. 2.5.11′(1), with a^{-1} in the role of a, and U in the role of W, we have

$$h_\mu(a^{-1}) \geq \log J(a, U).$$

From Cor. 5.5.2 and Lem. 2.5.11′(2), we have

$$h_\mu(a) \le \log J(a^{-1}, \widetilde{S}_-).$$

Combining these two inequalities with the fact that $h_\mu(a) = h_\mu(a^{-1})$ (see Exer. 2.3#7), we have

$$\log J(a, U) \le h_\mu(a^{-1}) = h_\mu(a) \le \log J(a^{-1}, \widetilde{S}_-).$$

Thus, if we show that

$$\log J(a^{-1}, \widetilde{S}_-) \le \log J(a, U), \tag{5.6.2}$$

then we must have equality throughout, and the desired conclusion will follow from Lem. 2.5.11′(3).

Because $\underline{u}$ belongs to the Lie algebra $\mathfrak{l}$ of L (see Notn. 5.3.3), the structure of $\mathfrak{sl}(2, \mathbb{R})$-modules implies, for each $\lambda \in \mathbb{Z}^+$, that the restriction $(\mathrm{ad}_\mathfrak{g}\,\underline{u})^\lambda|_{\mathfrak{g}_{-\lambda}}$ is a bijection from the weight space $\mathfrak{g}_{-\lambda}$ onto the weight space $\mathfrak{g}_\lambda$ (see Exer. 1). If $\underline{g} \in \widetilde{\mathfrak{s}}_- \cap \mathfrak{g}_{-\lambda}$, then Rem. 5.4.2(1) implies $\underline{g}(\mathrm{ad}_\mathfrak{g}\,\underline{u})^\lambda \in (\mathfrak{g}_- + \mathfrak{g}_0 + \mathfrak{u}) \cap \mathfrak{g}_\lambda = \mathfrak{u} \cap \mathfrak{g}_\lambda$, so we conclude that $(\mathrm{ad}_\mathfrak{g}\,\underline{u})^\lambda|_{\widetilde{\mathfrak{s}}_- \cap \mathfrak{g}_{-\lambda}}$ is an embedding of $\widetilde{\mathfrak{s}}_- \cap \mathfrak{g}_{-\lambda}$ into $\mathfrak{u} \cap \mathfrak{g}_\lambda$. So

$$\dim(\widetilde{\mathfrak{s}}_- \cap \mathfrak{g}_{-\lambda}) \le \dim(\mathfrak{u} \cap \mathfrak{g}_\lambda).$$

The eigenvalue of $\mathrm{Ad}_G\, a = \exp(\mathrm{ad}_\mathfrak{g}\,\underline{a})$ on $\mathfrak{g}_\lambda$ is e^λ, and the eigenvalue of $\mathrm{Ad}_G\, a^{-1}$ on $\mathfrak{g}_{-\lambda}$ is also e^λ (see Exer. 2). Hence,

$$
\begin{aligned}
\log J(a^{-1}, \widetilde{S}_-) &= \log \det(\mathrm{Ad}_G\, a^{-1})|_{\widetilde{\mathfrak{s}}_-} \\
&= \log \prod_{\lambda \in \mathbb{Z}^+} (e^\lambda)^{\dim(\widetilde{\mathfrak{s}}_- \cap \mathfrak{g}_{-\lambda})} \\
&= \sum_{\lambda \in \mathbb{Z}^+} (\dim(\widetilde{\mathfrak{s}}_- \cap \mathfrak{g}_{-\lambda})) \cdot \log e^\lambda \\
&\le \sum_{\lambda \in \mathbb{Z}^+} (\dim \mathfrak{u} \cap \mathfrak{g}_\lambda) \cdot \log e^\lambda \\
&= \log J(a, U),
\end{aligned}
$$

as desired. $\qquad\qquad\qquad\qquad\qquad\qquad\qquad\qquad\qquad\qquad\square$

Exercises for §5.6.

#1. Suppose
- $\mathcal{W}$ is a finite-dimensional real vector space, and
- $\rho\colon \mathfrak{sl}(2, \mathbb{R}) \to \mathfrak{sl}(\mathcal{W})$ is a Lie algebra homomorphism.

Show, for every $m \in \mathbb{Z}^{\ge 0}$, that $(\underline{u}^\rho)^m$ is a bijection from $\mathcal{W}_{-m}$ to $\mathcal{W}_m$.
[*Hint:* Use Prop. 4.9.22. If $-\lambda_i \le 2j - \lambda_i \le 0$, then $w_{i,j}(\underline{u}^\rho)^{\lambda_i - 2j}$

is a nonzero multiple of w_{i,λ_i-j}, and $w_{i,\lambda_i-j}(\underline{v}^\rho)^{\lambda_i-2j}$ is a nonzero multiple of $w_{i,j}$.]

#2. In the notation of the proof of Prop. 5.6.1, show that the eigenvalue of $\mathrm{Ad}_G\, a^{-1}$ on $\mathfrak{g}_{-\lambda}$ is the same as the the eigenvalue of $\mathrm{Ad}_G\, a$ on $\mathfrak{g}_\lambda$.

5.7. Completion of the proof

We wish to show, for some $x \in \Gamma\backslash G$, that $\mu(xS) > 0$. In other words, that $\mu(xS_-S_0S_+) > 0$. The following weaker result is a crucial step in this direction.

(5.7.1) Proposition. *For some $x \in \Gamma\backslash G$, we have $\mu(xS_-G_0G_+) > 0$.*

Proof. Assume that the desired conclusion fails. (This will lead to a contradiction.) Let Ω_ϵ be as in Cor. 5.5.4, with ϵ sufficiently small.

Because the conclusion of the proposition is assumed to fail, there exist $x, y \in \Omega_\epsilon$, with $x \approx y$ and $y = xg$, such that $g \notin S_-G_0G_+$. (See Exer. 2 for a non-infinitesimal proof.) Thus, we may write

$$g = vwh \text{ with } v \in S_-,\ w \in (G_- \ominus S_-) \smallsetminus \{e\}, \text{ and } h \in G_0G_+.$$

For simplicity, let us pretend that Ω_ϵ is S_--invariant. (This is not so far from the truth, because μ is S_--invariant and $\mu(\Omega_\epsilon)$ is very close to 1, so the actual proof is only a little more complicated (see Exer. 1).) Then we may replace x with xv, so that $g = wh \in (G_- \ominus S_-)G_0G_+$. This contradicts the definition of Ω_ϵ. $\square$

We can now complete the proof (using some of the theory of algebraic groups).

(5.7.2) Theorem. *μ is supported on a single S-orbit.*

Proof. There is no harm in assuming that G is almost Zariski closed (see Rem. 5.0.5). By induction on $\dim G$, we may assume that there does **not** exist a subgroup H of G, such that

- H is almost Zariski closed,
- $U \subset H$, and
- some H-orbit has full measure.

Then a short argument (see Exer. 4.7#8) implies, for all $x \in \Gamma\backslash G$, that

$$\text{if } V \text{ is any subset of } G,$$
$$\text{such that } \mu(xV) > 0, \text{ then } G \subset \overline{\overline{V}}. \tag{5.7.3}$$

This hypothesis will allow us to show that $S = G$.

Claim. We have $S_- = G_-$. Prop. 5.7.1 states that $\mu(xS_-G_0G_+) > 0$, so, from (5.7.3), we know that $G \subset \overline{S_-G_0G_+}$. This implies that $S_-G_0G_+$ must contain an open subset of G (see Exer. 5.4#6). Therefore

$$\dim S_- \geq \dim G - \dim(G_0G_+) = \dim G_-.$$

Because $S_- \subset G_-$, and G_- is connected, this implies that $S_- = G_-$, as desired.

The subgroup G_- is a horospherical subgroup of G (see Rem. 2.5.6), so we have shown that μ is invariant under a horospherical subgroup of G.

There are now at least three ways to complete the argument.

a) We showed that μ is G_--invariant. By going through the same argument, but with v^r in the place of u^t, we could show that μ is G_+-invariant. So S contains $\langle L, G_+, G_- \rangle$, which is easily seen to be a (unimodular) normal subgroup of G (see Exer. 3). Then Exer. 5.2#3 applies.

b) By using considerations of entropy, much as in the proof of Prop. 5.6.1, one can show that $G_+ \subset S$ (see Exer. 4), and then Exer. 5.2#3 applies, once again.

c) If we assume that $\Gamma \backslash G$ is compact (and in some other cases), then a completely separate proof of the theorem is known for measures that are invariant under a horospherical subgroup. (An example of an argument of this type appears in Exers. 3.2#8 and 5.) Such special cases were known several years before the general theorem. □

Exercises for §5.7.

#1. Prove Prop. 5.7.1 (without assuming Ω_ϵ is S_--invariant).
[*Hint:* Because Ω_ϵ contains 99% of the S_--orbits of both x and y, it is possible to find $x' \in xS_- \cap \Omega_\epsilon$ and $y' \in yS_- \cap \Omega_\epsilon$, such that $y' \in x'(G_- \ominus S_-)G_0G_+$.]

#2. Prove Prop. 5.7.1 without using infinitesimals.
[*Hint:* Use Cor. 5.5.4'.]

#3. Show that $\langle G_-, G_+ \rangle$ is a normal subgroup of G°.
[*Hint:* It suffices to show that it is normalized by G_-, G_0, and G_+.]

#4. (a) Show that $J(a^{-1}, G_-) = J(a, G_+)$.
   (b) Use Lem. 2.5.11' (at the beginning of §5.6) to show that if μ is G_--invariant, then it is G_+-invariant.

#5. Let
   • G be a connected, semisimple subgroup of $SL(\ell, \mathbb{R})$,
   • Γ be a lattice in G, such that $\Gamma \backslash G$ is compact,

- μ be a probability measure on $\Gamma \backslash G$,
- a^s be a nontrivial hyperbolic one-parameter subgroup of G, and
- G_+ be the corresponding expanding horospherical subgroup of G.

Show that if μ is G_+-invariant, then μ is the Haar measure on $\Gamma \backslash G$.

[*Hint:* Cf. hint to Exer. 3.2#8. (Let $U_\epsilon \subset G_+$, $A_\epsilon \subset G_0$, and $V_\epsilon \subset G_-$.) Because μ is not assumed to be a^s-invariant, it may not be possible to choose a generic point y for μ, such that $y a^{s_k} \to y$. Instead, show that the mixing property (3.2.8) can be strengthened to apply to the compact family of subsets $\{ y U_\epsilon A_\epsilon V_\epsilon \mid y \in \Gamma \backslash G \}$.]

5.8. Some precise statements

Let us now state these results more precisely, beginning with the statement that polynomials stay near their largest value for a proportional length of time.

(5.8.1) **Lemma.** *For any d and ϵ, and any averaging sequence $\{E_n\}$ of open sets in any unipotent subgroup U of G, there is a ball B around e in U, such that if*

- *$f: U \to \mathbb{R}^m$ is any polynomial of degree $\leq d$,*
- *E_n is an averaging set in the averaging sequence $\{E_n\}$, and*
- *$\sup_{u \in E_n} \|f(u)\| \leq 1$,*

then $\|f(v_1 u v_2) - f(u)\| < \epsilon$, for all $u \in E_n$, and all $v_1, v_2 \in B_n = a^{-n} B a^n$.

(5.8.2) **Remark.** Note that $v_U(B_n)/v_U(E_n) = v_U(B)/v_U(E)$ is independent of n; thus, B_n represents an amount of time proportional to E_n.

Proof. The set $\mathcal{P}^d$ of real polynomials of degree $\leq d$ is a finite-dimensional vector space, so

$$\left\{ f \in \mathcal{P}^d \ \middle| \ \sup_{u \in E} \|f(u)\| \leq 1 \right\}$$

is compact. Thus, there is a ball B around e in U, such that the conclusion of the lemma holds for $n = 0$. Rescaling by a^n then implies that it must also hold for any n. $\square$

As was noted in the previous chapter, if $y = xg$, then the relative displacement between xu and yu is $u^{-1}gu$. For each fixed g, this is a polynomial function on U, and the degree is bounded independent of g. The following observation makes a similar statement about the transverse divergence of two U-orbits. It is a formalization of Lem. 5.2.2.

(5.8.3) **Remark.** Given $y = xg$, the relative displacement between xu and yu is $u^{-1}gu$. To measure the part of this displacement that is transverse to the U-orbit, we wish to multiply by an element u' of U, to make $(u^{-1}gu)u'$ as small as possible: equivalently, we can simply think of $u^{-1}gu$ in the quotient space G/U. That is,

> *the transverse distance between the two U-orbits (at the point xu) is measured by the position of the point $(u^{-1}gu)U = u^{-1}gU$ in the homogeneous space G/U.*

Because U is Zariski closed (see Prop. 4.6.2), we know, from Chevalley's Theorem (4.5.2), that, for some m, there is

- a polynomial homomorphism $\rho \colon G \to \mathrm{SL}(m, \mathbb{R})$, and
- a vector $w \in \mathbb{R}^m$,

such that (writing our linear transformations on the left) we have

$$U = \{\, u \in G \mid \rho(u)w = w \,\}.$$

Thus, we may identify G/U with the orbit wG, and, because ρ is a polynomial, we know that $u \mapsto \rho(u^{-1}gu)w$ is a polynomial function on U. Hence, the transverse distance between the two U-orbits is completely described by a polynomial function.

We now make precise the statement in Lem. 5.2.1 that the direction of fastest divergence is in the direction of the normalizer. (See Rem. 5.8.5 for a non-infinitesimal version of the result.)

(5.8.4) **Proposition.** *Suppose*

- *U is a connected, unipotent subgroup of G,*
- *$x, y \in \Gamma \backslash G$,*
- *$y = xg$, for some $g \in G$, with $g \approx e$, and*
- *E is an (infinitely large) averaging set,*

such that

- *$g^E = \{\, u^{-1}gu \mid u \in E \,\}$ has finite diameter in G/U.*

Then each element of $g^E U/U$ is infinitesimally close to some element of $N_G(U)/U$.

Proof. Let $g' \in g^E$. Note that

$$N_G(U)/U = \{\, x \in G/U \mid ux = x,\ \text{for all } u \in U \,\}.$$

Thus, it suffices to show that $ug'U \approx g'U$, for each finite $u \in U$.

We may assume $g'U$ is a finite (not infinitesimal) distance from the base point eU, so its distance is comparable to the far-thest distance in $g^E U/U$. It took infinitely long to achieve this distance, so polynomial divergence implies that it takes a pro-portional, hence infinite, amount of time to move any additional finite distance. Thus, in any finite time, the point $g'U$ moves only infinitesimally. Therefore, $ug'U \approx g'U$, as desired.   □

(5.8.5) **Remark.** The above statement and proof are written in terms of infinitesimals. To obtain a non-infinitesimal version, replace

- x and y with convergent sequences $\{x_k\}$ and $\{y_k\}$, such that $d(x_k, y_k) \to e$,
- g with the sequence $\{g_k\}$, defined by $x_k g_k = y_k$, and
- E with an averaging sequence E_n, such that $g_k^{E_{n_k}}$ is bounded in G/U (independent of k).

The conclusion is that if $\{g'_k\}$ is any sequence, such that

- $g'_k \in g_k^{E_{n_k}}$ for each k, and
- $g'_k U/U$ converges,

then the limit is an element of $N_G(U)/U$ (see Exer. 3).

(5.8.6) **Lemma.** *If*

- *C is any compact subset of $N_G(U) \smallsetminus \mathrm{Stab}_G(\mu)$, and*
- *$\epsilon > 0$,*

then there is a compact subset K of $\Gamma \backslash G$, such that

1) *$\mu(K) > 1 - \epsilon$ and*
2) *$K \cap Kc = \varnothing$, for all $c \in C$.*

Proof. Let Ω be the set of all points in $\Gamma \backslash G$ that are generic for μ (see Defn. 3.1.5 and Thm. 3.4.3). It suffices to show that $\Omega \cap \Omega c = \varnothing$, for all $c \in N_G(U) \smallsetminus \mathrm{Stab}_G(\mu)$, for then we may choose K to be any compact subset of Ω with $\mu(K) > 1 - \epsilon$.

Fix $c \in C$. We choose a compact subset K_c of Ω with $K_c \cap K_c c = \varnothing$, and $\mu(K_c) > 1 - \delta$, where δ depends only on c, but will be specified later.

Now suppose $x, xc \in \Omega$. Except for a proportion δ of the time, we have xu very near to K_c (because $x \in \Omega_c$). Thus, it suffices to have xuc very close to K_c more than a proportion δ of the time. That is, we wish to have $(xc)(c^{-1}uc)$ very close to K_c a significant proportion of the time.

We do have $(xc)u$ very close to K_c a huge proportion of the time. Now c acts on U by conjugation, and the Jacobian of this diffeomorphism is constant (hence bounded), as is the maximum eigenvalue of the derivative. Thus, we obtain the desired conclusion by choosing δ sufficiently small (and E to be a nice set) (see Exer. 4). □

(5.8.7) **Completing the proof of Prop. 5.2.4'.** Fix a set Ω_0 as in the Uniform Pointwise Ergodic Theorem (3.4.4). Suppose $x, y \in \Omega_0$ with $x \approx y$, and write $y = xg$. Given an (infinite) averaging set $E_n = a^{-n}Ea_n$, such that g^{E_n} is bounded in G/U, and any $v \in E$, we wish to show that $(a^{-n}va^n)gU$ is infinitesimally close to $\mathrm{Stab}_G(\mu)/U$. The proof of Prop. 5.2.4' will apply if we show that yu' is close to K a significant proportion of the time.

To do this, we make the additional technical assumption that

$$g^* = a^n g a^{-n} \text{ is finite (or infinitesimal).} \qquad (5.8.8)$$

Let us assume that E is a ball around e. Choose a small neighborhood B of v in E, and define

$$\sigma : B \to U \text{ by } ug^* \in (G \ominus U) \cdot \sigma(u), \qquad \text{for } u \in B,$$

so

$$u' = a^{-n} \sigma(u) a^n.$$

The Jacobian of σ is bounded (between $1/J$ and J, say), so we can choose ϵ so small that

$$(1 - J^2 \epsilon) \cdot \nu_U(B) > \epsilon \cdot \nu_U(E).$$

(The compact set K should be chosen with $\mu(K) > 1 - \epsilon$.)

By applying Cor. 3.4.4 to the averaging sequence $\sigma(B)_n$ (and noting that n is infinitely large), and observing that

$$y(a^{-n}ua^n) = x(a^{-n}g^*ua^n),$$

we see that

$$\nu_U(\{ u \in \sigma(B) \mid x(a^{-n}g^*ua^n) \not\approx K \}) \le \epsilon \, \nu_U(B_n).$$

Therefore, the choice of ϵ implies

$$\nu_U(\{ u \in B \mid x(a^{-n}g^* \sigma(u) a^n) \approx K \}) > \epsilon \, \nu_U(E).$$

Because

$$\begin{aligned}
x(a^{-n}g^* \sigma(u) a^n) &= x(a^{-n}g^*a^n)(a^{-n}\sigma(u) a^n) \\
&= xgu' \\
&= yu',
\end{aligned}$$

this completes the proof. □

(5.8.9) Technical assumption.

1) The technical assumption (5.8.8) in the proof of Prop. 5.2.4′ can be stated in the following explicit form if g is infinitesimal: there are
   - an (infinite) integer n, and
   - a finite element u_0 of U,

   such that

   (a) $a^{-n} u_0 a^n g \in G_- G_0 G_+$,

   (b) $a^{-n} u_0 a^n g U$ is not infinitesimally close to eU in G/U, and

   (c) $a^n g a^{-n}$ is finite (or infinitesimal).

2) In non-infinitesimal terms, the assumption on $\{g_k\}$ is: there are
   - a sequence $n_k \to \infty$, and
   - a bounded sequence $\{u_k\}$ in U,

   such that

   (a) $a^{-n_k} u_k a^{n_k} g_k \in G_- G_0 G_+$,

   (b) no subsequence of $a^{-n_k} u_k a^{n_k} g_k U$ converges to eU in G/U, and

   (c) $a^{n_k} g_k a^{-n_k}$ is bounded.

Exercises for §5.8.

#1. Show that if $g = a^{-t} v a^t$, for some standard $v \in G_-$, and g is infinitesimal, then either

   (a) g satisfies the technical assumption (5.8.9), or

   (b) $g \in \widetilde{S}_-$.

#2. Show that if g is as in Lem. 5.5.3, then g satisfies the technical assumption (5.8.9).

   [*Hint:* Choose $n > 0$ so that $a^n g a^{-n}$ is finite, but not infinitesimal. Then $a^n g a^{-n}$ is not infinitesimally close to $\widetilde{S}$, so there is some (small) $u \in U$, such that $u(a^n g a^{-n})$ is not infinitesimally close to $G_- G_0 U$. Conjugate by a^n.]

#3. Provide a (non-infinitesimal) proof of Rem. 5.8.5.

#4. Complete the proof of Lem. 5.8.6, by showing that if E is a convex neighborhood of e in U, and δ is sufficiently small, then, for all n and every subset X of E_n with $\nu_U(X) \geq (1 - \delta)\nu_U(E_n)$, we have

$$\nu_U(\{ u \in E_n \mid c^{-1} u c \in X \}) > \delta \, \nu_U(X).$$

[*Hint:* There is some $k > 0$, such that

$$c^{-1}E_{n-k}c \subset E_n, \text{ for all } n.$$

Choose δ small enough that

$$\nu_U(E_{n-k}) > (J+1)\delta\nu_U(E_n),$$

where J is the Jacobian of the conjugation diffeomorphism.]

#5. Prove the non-infinitesimal version of Prop. 5.2.4′: For any $\epsilon > 0$, there is a compact subset Ω_ϵ of $\Gamma\backslash G$, with $\mu(\Omega_\epsilon) > 1 - \epsilon$, and such that if
   - $\{x_k\}$ and $\{y_k\}$ are convergent sequences in Ω_ϵ,
   - $\{g_k\}$ is a sequence in G that satisfies 5.8.9(2),
   - $x_k g_k = y_k$,
   - $g_k \to e$,
   - $\{E_n\}$ is an averaging sequence, and $\{n_k\}$ is a sequence of natural numbers, such that $g_k^{E_{n_k}}$ is bounded in G/U (independent of k),
   - $g'_k \in g_k^{E_{n_k}}$, and
   - $g'_k U/U$ converges,

   then the limit of $\{g'_k U\}$ is an element of S/U.

#6. Prove Cor. 5.5.2 without using infinitesimals.

#7. Prove the non-infinitesimal version of Lem. 5.5.3: If
   - $\{g_n\}$ is a sequence in $(G_- \ominus \widetilde{S}_-)G_0G_+$,
   - $\{E_n\}$ is an averaging sequence, and $\{n_k\}$ is a sequence of natural numbers, such that $g_k^{E_{n_k}}$ is bounded in G/U (independent of k),
   - $g'_k \in g_k^{E_{n_k}}$, and
   - $g'_k U/U$ converges,

   then the limit of $\{g'_k U\}$ is an element of G_+/U.

#8. Prove Cor. 5.5.4′.

5.9. How to eliminate Assumption 5.3.1

Let $\hat{U}$ be a maximal connected, unipotent subgroup of S, and assume $\{u^t\} \subset \hat{U}$.

From Rem. 5.8.3, we know, for $x, y \in \Gamma\backslash G$ with $x \approx y$, that the transverse component of the relative position between xu and yu is a polynomial function of u. (Actually, it is a rational function (cf. Exer. 1d), but this technical issue does not cause any serious problems, because the function is unbounded on U, just like a polynomial would be.) Furthermore, the transverse component belongs to S (usually) and normalizes $\hat{U}$ (see Props. 5.2.4′ and 5.8.4). Let $\hat{S}$ be the closure of the subgroup of $N_S(\hat{U})$ that is generated by the image of one of these polynomial maps, together with $\hat{U}$. Then $\hat{S}$ is (almost) Zariski closed (see Lem. 4.6.9), and the maximal unipotent subgroup $\hat{U}$ is normal, so the structure theory of algebraic groups implies that there is a hyperbolic torus T of $\hat{S}$, and a compact subgroup C of $\hat{S}$, such that $\hat{S} = TC\hat{U}$ (see Thm. 4.4.7 and Cor. 4.4.10(3)). Any nonconstant polynomial is unbounded, so (by definition of $\hat{S}$), we see that $\hat{S}/\hat{U}$ is not compact; thus, T is not compact. Let

- $\{a^s\}$ be a noncompact one-dimensional subgroup of T, and
- $U = S_+$.

This does not establish (5.3.1), but it comes close:

- μ is invariant under $\{a^s\}$, and
- $\{a^s\}$ is hyperbolic, and normalizes U.

We have not constructed a subgroup L, isomorphic to $\mathrm{SL}(2, \mathbb{R})$, that contains $\{a^s\}$, but the only real use of that assumption was to prove that $J(a^{-s}, \widetilde{S}_-) \leq J(a^s, S_+)$ (see 5.6.2). Instead of using the theory of $\mathrm{SL}(2, \mathbb{R})$-modules, one shows, by using the theory of algebraic groups, and choosing $\{a^s\}$ carefully, that $J(a^s, H) \geq 1$, for every Zariski closed subgroup H of G that is normalized by $a^s U$ (see Exer. 2).

An additional complication comes from the fact that a^s may not act ergodically (w.r.t. μ): although u^t is ergodic, we cannot apply the Mautner Phenomenon, because $U = S_+$ may not contain $\{u^t\}$ (since $(u^t)_-$ or $(u^t)_0$ may be nontrivial). Thus, one works with ergodic components of μ. The key point is that the arguments establishing Prop. 5.6.1 actually show that each ergodic component of μ is $\widetilde{S}_-$-invariant. But then it immediately follows that μ itself is $\widetilde{S}_-$-invariant, as desired, so nothing was lost.

Exercises for §5.9.

#1. Let $B = \left\{ \begin{bmatrix} * & * \\ 0 & * \end{bmatrix} \right\} \subset \mathrm{SL}(2, \mathbb{R})$, and define $\psi \colon \mathbb{U}_2 \times B \to \mathrm{SL}(2, \mathbb{R})$ by $\psi(u, b) = ub$.

    (a) Show ψ is a polynomial.

    (b) Show ψ is injective.

    (c) Show the image of ψ is a dense, open subset $\mathcal{O}$ of $\mathrm{SL}(2, \mathbb{R})$.

(d) Show ψ^{-1} is a rational function on $\mathbb{O}$.

[*Hint:* Solve $\begin{bmatrix} x & y \\ z & w \end{bmatrix} = \begin{bmatrix} 1 & 0 \\ u & 1 \end{bmatrix} \begin{bmatrix} a & b \\ 0 & 1/a \end{bmatrix}$ for a, b, and u.]

#2. Show there is a (nontrivial) hyperbolic one-parameter subgroup $\{a^s\}$ of $\hat{S}$, such that $J(a^s, H) \geq 1$, for every almost-Zariski closed subgroup H of G that is normalized by $\{a^s\} U$, and every $s > 0$.

[*Hint:* Let $\phi\colon U \to \hat{S}$ be a polynomial, such that $\langle \phi(U), U \rangle = \hat{S}$. For each H, we have $J(u, H) = 1$ for all $u \in U$, and the function $J(\phi(u), H)$ is a polynomial on U. Although there may be infinitely many different possibilities for H, they give rise to only finitely many different polynomials, up to a bounded error. Choose $u \in U$, such that $|J(\phi(u), H)|$ is large for all H, and let $a^1 = \phi(u)_h$ be the hyperbolic part in the Jordan decomposition of $\phi(u)$.]

Notes

Our presentation in this chapter borrows heavily from the original proof of M. Ratner [2, 3, 4], but its structure is based on the approach of G. A. Margulis and G. M. Tomanov [1]. The two approaches are similar at the start, but, instead of employing the entropy calculations of §5.6 to finish the proof, Ratner [3, Lem. 4.1, Lem. 5.2, and proof of Lem. 6.2] bounded the number of small rectangular boxes needed to cover certain subsets of $\Gamma \backslash G$. This allowed her to show [3, Thm. 6.1] that the measure μ is supported on an orbit of a subgroup L, such that

- L contains both u^t and a^s,
- the Jacobian $J(a^s, \mathfrak{l})$ of a^s on the Lie algebra of L is 1, and
- $L_0 L_+ \subset \mathrm{Stab}_G(\mu)$.

Then an elementary argument [3, §7] shows that $L \subset \mathrm{Stab}_G(\mu)$.

The proof of Margulis and Tomanov is shorter, but less elementary, because it uses more of the theory of algebraic groups.

§5.2. Lemma 5.2.1 is a version of [2, Thm. 3.1 ("R-property")] and (the first part of) [1, Prop. 6.1].

Lemma 5.2.2 is implicit in [2, Thm. 3.1] and is the topic of [1, §5.4].

Proposition 5.2.4′ is [1, Lem. 7.5]. It is also implicit in the work of M. Ratner (see, for example, [3, Lem. 3.3]).

§5.4. The definition (5.4.1) of $\tilde{S}$ is based on [1, §8.1] (where $\tilde{S}$ is denoted $\mathcal{F}(s)$ and $\tilde{S}_-$ is denoted $U^-(s)$).

§5.5. Corollary 5.5.2 is [1, Cor. 8.4].

Lemma 5.5.3 is a special case of the last sentence of [1, Prop. 6.7].

§5.6. Proposition 5.6.1 is [1, Step 1 of 10.5].

§5.7. The proof of Prop. 5.7.1 is based on [1, Lem. 3.3].

The *Claim* in the proof of Thm. 5.7.2 is [1, Step 2 of 10.5].

The use of entropy to prove that if μ is G_--invariant, then it is G_+-invariant (alternative (b) on p. 183) is due to Margulis and Tomanov [1, Step 3 of 10.5].

References to results on invariant measures for horospherical subgroups (alternative (c) on p. 183) can be found in the historical notes at the end of Chap. 1.

Exercise 4.7#8 is [1, Prop. 3.2].

§5.8. The technical assumption (5.8.9) needed for the proof of (5.2.4') is based on the condition (∗) of [1, Defn. 6.6]. (In Ratner's approach, this role is played by [3, Lem. 3.1] and related results.)

Exers. 5.8#1 and 5.8#2 are special cases of [1, Prop. 6.7].

§5.9. That $\hat{S}/U$ is not compact is part of [1, Prop. 6.1].

That a^s may be chosen to satisfy the condition $J(a^s, H) \geq 1$ is [1, Prop. 6.3b].

The (possible) nonergodicity of μ is addressed in [1, Step 1 of 10.5].

References

[1] G. A. Margulis and G. M. Tomanov: Invariant measures for actions of unipotent groups over local fields on homogeneous spaces. *Invent. Math.* 116 (1994) 347–392. (Announced in *C. R. Acad. Sci. Paris Sér. I Math.* 315 (1992), no. 12, 1221–1226.) MR 94f:22016, MR 95k:22013

[2] M. Ratner: Strict measure rigidity for unipotent subgroups of solvable groups. *Invent. Math.* 101 (1990), 449–482. MR 92h:22015

[3] M. Ratner: On measure rigidity of unipotent subgroups of semisimple groups. *Acta Math.* 165 (1990), 229–309. MR 91m:57031

[4] M. Ratner: On Raghunathan's measure conjecture. *Ann. of Math.* 134 (1991), 545–607. MR 93a:22009

List of Notation

Chapter 1. Introduction to Ratner's Theorems

$\mathbb{T}^n = \mathbb{R}^n / \mathbb{Z}^n = n$-torus, 1

G = Lie group, 2

Γ = lattice in G, 2

u^t = unipotent one-parameter subgroup, 2

$\varphi_t = u^t$-flow on $\Gamma \backslash G$, 2

$[x]$ = image of x in $\Gamma \backslash G$, 2

$\Gamma \backslash G = \{ \Gamma x \mid x \in G \}$, 2

$SL(2, \mathbb{R})$ = group of 2×2 real matrices of determinant one, 3

$u^t = \begin{bmatrix} 1 & 0 \\ t & 1 \end{bmatrix}$, 3

$a^t = \begin{bmatrix} e^t & 0 \\ 0 & e^{-t} \end{bmatrix}$, 3

η_t = horocycle flow = u^t-flow on $\Gamma \backslash SL(2, \mathbb{R})$, 3

γ_t = geodesic flow = a^t-flow on $\Gamma \backslash SL(2, \mathbb{R})$, 3

μ = measure on G or on $\Gamma \backslash G$, 4

$\mathcal{F}$ = fundamental domain for Γ in G, 4

μ_G = G-invariant ("Haar") probability measure on $\Gamma \backslash G$, 4

$\mathfrak{H}$ = hyperbolic plane, 8

$\mathrm{Stab}_H(\Gamma x)$ = stabilizer of Γx in H, 9

$SO(Q)$ = orthogonal group of quadratic form Q, 14

$SO(m, n)$ = orthogonal group of quadratic form $Q_{m,n}$, 14

$Q_{m,n}$ = quadratic form $x_1^2 + \cdots + x_m^2 - x_{m+1}^2 - \cdots - x_{m+n}^2$, 14

H° = identity component of H, 15

$v^\perp$ = orthogonal complement of vector v, 17

μ_S = S-invariant probability measure on an S-orbit, 21

$d(x, y)$ = distance from x to y, 34

I = identity matrix, 34

Chapter 2. Introduction to Entropy

Chapter 3. Facts from Ergodic Theory

Chapter 4. Facts about Algebraic Groups

$\mathbb{R}[x_{1,1}, \ldots, x_{\ell,\ell}]$ = real polynomials in $\{x_{i,j}\}$, 123

$g_{i,j}$ = entries of matrix g, 123

Q = some subset of $\mathbb{R}[x_{1,1}, \ldots, x_{\ell,\ell}]$, 123

$\mathrm{Var}(Q)$ = variety associated to set Q of polynomials, 123

$\mathbb{D}_\ell$ = {diagonal matrices in $\mathrm{SL}(\ell, \mathbb{R})$}, 124

$\mathbb{U}_\ell$ = {unipotent lower-triangular matrices in $\mathrm{SL}(\ell, \mathbb{R})$}, 124

$\mathrm{Stab}_{\mathrm{SL}(\ell,\mathbb{R})}(v)$ = stabilizer of vector v, 124

$\mathrm{Stab}_{\mathrm{SL}(\ell,\mathbb{R})}(V)$ = stabilizer of subspace V, 124

$\mathcal{I}(Z)$ = ideal of polynomials vanishing on Z, 126

$\overline{\overline{H}}$ = Zariski closure of H, 127

g_u, g_h, g_e = Jordan components of matrix g, 131

$\mathrm{trace}\, g$ = trace of the matrix g, 133

$A \ltimes B$ = semidirect product, 136

$g^\rho = \rho(g)$, 140

$\mathbb{R}\mathrm{P}^{m-1}$ = real projective space, 147

$[v]$ = image of vector v in $\mathbb{R}\mathrm{P}^{m-1}$, 147

$\mathrm{Mat}_{\ell \times \ell}(\mathbb{R})$ = all real $\ell \times \ell$ matrices, 155

$\mathfrak{g}, \mathfrak{h}, \mathfrak{u}, \mathfrak{s}, \mathfrak{l}$ = Lie algebra of G, H, U, S, L, 155

$[\underline{x}, \underline{y}] = \underline{x}\underline{y} - \underline{y}\underline{x}$ = Lie bracket of matrices, 155

$\mathrm{Rad}\, G$ = radical of the Lie group G, 157

Ad_G = adjoint representation of G on its Lie algebra, 157

$\mathrm{ad}_\mathfrak{g}$ = adjoint representation of the Lie algebra $\mathfrak{g}$, 158

Chapter 5. Proof of the Measure-Classification Theorem

μ = ergodic u^t-invariant probability measure on $\Gamma \backslash G$, 165

$S = \mathrm{Stab}_G(\mu)^\circ$, 169

L = subgroup locally isomorphic to $\mathrm{SL}(2, \mathbb{R})$ containing u^t, 171

a^s = hyperbolic one-parameter subgroup in $\mathrm{Stab}_{N_L(\{u^t\})}(\mu)$, 171

$\mathfrak{g}, \mathfrak{s}, \mathfrak{l}$ = Lie algebra of G, S, L, 172

$\underline{g} = \log g$, 172

v^r = unipotent subgroup of L opposite to $\{u^t\}$, 172

$\mathfrak{g}_-, \mathfrak{g}_0, \mathfrak{g}_+$ = subspaces of $\mathfrak{g}$ (determined by a^s), 172

$\mathfrak{s}_-, \mathfrak{s}_0, \mathfrak{s}_+$ = subspaces of $\mathfrak{s}$ (determined by a^s), 172

$G_-, G_0, G_+, S_-, S_0, S_+$ = corresponding subgroups, 172

$U = S_+$, 172

$\mathfrak{u}$ = Lie algebra of U, 172

$\tilde{S} = \left\{ g \in G \mid u^{-1}gu \in \overline{G_-G_0U}, \text{ for all } u \in U \right\}$, 173

$\tilde{S}_- = \tilde{S} \cap G_-$, 173

$\mathfrak{g}_+ \ominus \mathfrak{u} = \text{complement to } \mathfrak{u} \text{ in } \mathfrak{g}_+$, 178

$\mathfrak{g}_- \ominus \tilde{\mathfrak{s}}_- = \text{complement to } \tilde{\mathfrak{s}}_- \text{ in } \mathfrak{g}_-$, 178

$G_+ \ominus U = \exp(\mathfrak{g}_+ \ominus \mathfrak{u})$, 178

$G_- \ominus \tilde{S}_- = \exp(\mathfrak{g}_- \ominus \tilde{\mathfrak{s}}_-)$, 178

Index

adjoint
  of an operator, 110
  representation, *see*
    representation, adjoint
algebraic closure, 153
algebraic group
  compact, 137, 138, 142
  defined over $\mathbb{Q}$, 16, 151-154
  over $\mathbb{C}$, 160
  over $\mathbb{R}$, 123
  real, 15, 16, 123-154, 160
  simply connected, 162
  theory of, 16, 123, 157, 160-162,
    166, 182, 190, 191
  unipotent, *see*
    unipotent subgroup
atom (of a partition), 80, 84, 87,
  88, 93, 94, 96, 97, 99
automorphism, 45
  expanding, 116-118
  inner, 45
averaging
  sequence, 116-119, 184-189
  set, 116, 185, 187

babysitting, 37
Baker's Transformation, 75, 76, 78,
  79, 86, 89, 90, 100
Banach space, 114
basis (of a vector space), 28
Bernoulli shift, 32, 65, 73-78, 86,
  88, 100
bilinear form
  nondegenerate, 17
  symmetric, 17
bit (of information), 80
butterfly flapping its wings, 77

Cantor set, 4
Cartan Decomposition, 111
centralizer, 48, 49, 130, 132, 137
chain condition
  ascending, 126, 127
  descending, 126, 127
Claim, 43, 192
classical limit, 28
coin tossing, 73, 78, 82
  history, 74
commutation relations, 3, 8
commutative algebra, 152
complement (of a subspace), 159,
  178, 180
component
  connected, 7, 15, 18, 60, 88, 125,
    128, 129, 145, 150, 166
  ergodic, *see* ergodic component
  G_+-, 167, 173
  identity, 15, 51, 128, 135, 174
  irreducible, 126, 127
  Jordan, 132
  of an ordered pair, 57
  transverse, 49, 189, 190
continued fraction, 64
converge uniformly, 46, 47
convex
  combination, 114, 115
  set, 114
coset
  left, 2
  right, 2
countable-to-one, 40
counterexample, 3, 14, 23, 26, 31,
  32, 176
counting rectangular boxes, 191

dense orbit, 3, 4, 6, 19, 79